Dictionary

Of

Explosions & Explosives

Second Edition

Nurul Alam, Chief Inspector of
Explosives in Bangladesh

Nurul Alam

Dictionary of Explosions & Explosives

Second Edition

International Society of Explosives Engineers
Cleveland, Ohio ▪ USA

Published by
International Society of Explosives Engineers
30325 Bainbridge Road
Cleveland, Ohio USA 44139-2295
www.isee.org

LCCN: 20079244738

ISBN: 1-892396-16-5

Printed in the United States of America

Dedicated to my
motherly daughter Nabila
and
affectionate son Asif
whose enthusiastic
inspiration and helpfulness
have made writing of
the book a pleasant task.

Preface

'Nothing in life is to be feared; It is only to be understood.'

Keeping the great plea of Madame Curie in mind I have made a humble attempt to an understanding of the complex terms of explosions and explosives in a simpler way.

An understanding of the factors that trigger an explosion and the material which could explode enables a person to avert the unwanted and tragic situation that may arise from an explosion. Better understanding can also help face the emergencies. But unlike other fields of scientific knowledge explosions and explosives are inadequately understood.

Besides, in today's society, there is hardly any item that is not touched in its extraction or processing by some form of energy requiring the use of commercial or industrial explosives. An inquisitive reader may be interested to have knowledge about explosives. With this end in view, most of the commonly encountered terms of explosions and explosives have been listed in this book. Feature articles on some important terms and topics have also been included.

It is hoped that this dictionary will be an ideal reference tool for Safety Executives, Safety Engineers, Explosives Engineers, Fire Officers, Students and Scientists. There are elements of interest for inquisitive general readers.

Literature on the subject is very wide. A claim for completeness is therefore absurd. Yet an earnest effort has been made to satisfy the queries of professionals including Defense Service Officials, Police Personnel, and Emergency Response Professionals in their battle against terrorist acts.

A good number of pertinent terms on fire, firearm, chemical warfare agent, safety etc. as incorporated in this book will be an extra attraction of the reader.
As is usual, one may not know about various types of, say, explosion; hence other types thereof have been listed below the defining term 'explosion'.

Some of the older explosives and compositions have also been included in order to give a picture of the gradual development in the particular field or because they are considered to be of particular historic interest.

In my long professional life as a chemist and as an enthusiastic safety executive in the field of explosives, pressure vessels and flammables I have consulted a good number of books and journals, the titles, authors or publishers of each such book is neither in my memory nor in my reach. But I have gained humble knowledge from such books that actually helped me compile this dictionary. However, the titles, authors or publishers of the books and the journals which are in my possession, record or memory are cited in bibliography at the end this book.

Finally, I acknowledge with thanks and gratitude the help and assistance I have received from Mr. Zainal Abedin, Mr. A.S.M. Ashraful Huq, Syed Sayem Ahmed, Syed R. Kabir, my junior colleagues in the Department of Explosives, especially Mr. Khairul Basher, Mr. Shamsul Alam, Mr. Farid uddin Ahmed, Mr. A. Rab & Mr. A. Matin and my eldest brother A. Satter Hanafee. I would also like to thank my wife Syeda Selina Begum for her support during the many months of frantic efforts in writing the book.

Dhaka
March 26, 2005.

Nurul Alam
Chief Inspector of Explosives in Bangladesh.

Preface to the Revised Second Edition.

As was expected, the Dictionary of Explosions and Explosives has been appreciated by expert personalities to be an excellent reference tool. Patronage from the International Society of Explosives Engineers (ISEE) has made the book available globally to interested readers. These prompted to publish the revised second edition.

A fairly good number of new items have been added. A further attempt has been made to make the Dictionary a guide for students and trainees. Some minor typographical mistakes that had been in the first edition have also been corrected.

I should like to express my sincere thanks and gratitude to Mr. Jeffrey L. Dean, CAE, Executive Director of the International Society of Explosives Engineers for his generous cooperation and help in promoting the book and publishing the revised second edition.

February, 2007. Nurul Alam.

Publisher's Note: Mr. Nurul Alam passed away in August 2007. The book was published posthumously in his memory.

Guidelines for the User of the Dictionary

1. Alphabetization

The terms in this dictionary are alphabetized on a letter-by-letter basis; comma, hyphen and word spacing in a term are ignored on the sequencing.

2. Cross-referencing

Abbreviations, acronyms, antonyms, and synonyms are added to the definitions and are also listed in the alphabetical sequence as cross-references to the defining terms.

A term that appears elsewhere in this dictionary is printed in *italic* under the defining term for the convenience of inquisitive readers.

AAR's Bureau of Explosives – The Bureau of Explosives (BOE) branch of the Association of American Railroads (AAR) serves to promote the safe transportation of explosives and other hazardous materials, and respond to emergencies in North America. The BOE, with its inspector force and publications, is a major and vital force in the safe transportation of explosives and other hazardous materials.

Abbcite – An earlier type of *ammonia dynamite* that contains a relatively high amount of alkali chloride. Sodium chloride is incorporated in the product with a view to reducing brisance heat of detonation in order to lower the risk of firedamp explosions. Hence used as a *permitted/permissible* blasting explosive in gassy coalmines.

Abel Heat Test – A prompt routine test for demonstrating the absence of impurities, in a sample of nitrocellulose, nitroglycerin and nitroglycol, causing low thermal stability.

Abelite – An explosive, named after the British chemist Sir Frederick Abel (died 1902), consisting essentially of *ammonium nitrate* and a nitro derivative of some aromatic hydrocarbon, e.g. *TNT*.

A-Black Powder – See *Powder, A-Black*.

A-Blasting Powder – See *Powder, A-Blasting*.

A-Bomb – See *Atom Bomb* and *Nuclear Weapon*.

Absolute Bulk Strength (ABS) – See *Explosive Energy.*

Absolute Weight Strength (AWS) – See *Explosive Energy.*

Acardite – Synonym of *Diphenylurea.*

Accelerant – Accelerant means anything, such as kerosene, which helps spread fire.

Acceptor – A charge of an explosive or a blasting agent accepting an impulse from an exploding donor charge.

Accessories, Blasting – See *Blasting Accessories.*

Accident, Major – See *Major Accident.*

Accuracy of Fire – Accuracy of fire is the measurement of precision of fire. It is expressed as the distance between the centers of impact to that of the target.

Acetylene – Acetylene, C_2H_2, HC≡CH, is a highly flammable gas with a very wide range between the *lower explosive limit* (*LEL*) of 2.5% by volume in air, and the *upper explosive limit* (*UEL*) of 82% by volume in air. Apart from this, if the pressure exceeds two atmospheres the gas explodes, even in the absence of air. Because of this property, acetylene is stored and transported in steel cylinders containing a solvent and a porous material.

It is a colorless gas which when pure has a pleasant smell. It is slightly lighter than air (air = 1, acetylene = 0.906). Its melting point is -84.7°C and boiling point is -80.7°C.

Although it may be manufactured by newer methods, such as the cracking of hydrocarbons in an electric arc, the pyrolysis of the lower paraffins in the presence of steam, and the partial oxidation of natural gas (methane), the original method for the manufacture of acetylene, the action of water on calcium carbide, is still of great importance.

Acetylene is widely used for oxy-acetylene welding. It was first used for lighting homes, marine buoys, railways, and mines. It has now become the starting point for the manufacture of a wide range of chemicals.

Synonym: Ethyne.

Acetylide – An acetylide is a high explosive composed of a metal combined with a carbon-to-carbon linkage, which is the explosive group, -C≡C-.

It is sometimes known as metallic carbide. Acetylene, under certain conditions, forms highly explosive metallic compounds such as acetylides of copper, silver, and mercury.

Acetylide is one of the few commercial explosives that contain no oxygen or nitrogen. It is very sensitive to shock, friction and heat, and hence it explodes instantaneously. The explosion of an acetylide produces no gas but simply is an effect of the large amount of heat produced readily.

An acetylide is used for detonating compositions, or in combination with lead azide in detonating rivets, where the acetylide reduces the flash point of the more insensitive azides.

Acetylide, Copper – See *Copper Acetylide.*

Acetylide, Silver – See *Silver Acetylide.*

Actuator – A mechanical device that is operated by a pressure cartridge. The pin puller is an example of an actuator.

Acremite – A blasting agent consisting of approximately 94% ammonium nitrate and 6% fuel oil, named by the U.S. inventor Acre. At the beginning it was prepared by the user in a primitive way for open cast mining in dry conditions.

A.D.C. Test – Acronym for Ardeer Double Cartridge Test. It is a test for measuring the ability of a *cartridge* of explosive to propagate over air gaps.

Adiabatic – Occurring without gain or loss of heat; a change of the properties, such as volume and pressure of the contents of an enclosure, without exchange of heat between the enclosure and its surroundings.

Adiabatic Flame Temperature – As applied to interior ballistics calculation, the temperature that the gaseous products of combustion of the propellant would attain if maintained at constant volume and without loss of energy to the surrounding medium.

Adiabatic Flame Temperature, Isobaric – Isobaric Adiabatic Flame Temperature: Adiabatic flame temperature attained under constant pressure conditions.

Adiabatic Flame Temperature, Isochoric – Isochoric adiabatic flame temperature: Adiabatic flame temperature attained under constant volume conditions.

Adiabatic Temperature – The temperature attained by a system undergoing a volume or pressure change in which no heat enters or leaves the system.

Adiabatic Bomb Calorimeter – Synonym of *Bomb Calorimeter.*

Adobe Charge – An explosive charge, either mud-covered or unconfined, fired in contact with a rock surface without making use of a borehole.

Synonyms: *Bulldoze* and *Mudcap.*

Advance – The term is applied to mining or tunneling. It means the distance by which the face of a tunnel is moved forward by each round of blasting.

Aerial Flare – An aerial flare is a pyrotechnic device, which is dropped from an aircraft and is designed to illuminate an area on the surface below the aircraft.

Aerial Shell – It is a device used in high-level displays. It may be either spherical or cylindrical. Such a device is propelled high in the air from a mortar, which is either buried in the ground or secured in a rack, trough, or drum. A shell generally ranges from 5 centimeters (2 inches) to 30 centimeters (12 inches) in diameter.

Aerobic Explosion – Aerobic explosion is a type of chemical explosion in which a chemical reaction occurs only in the presence of oxygen. When a combustible substance is intimately mixed with oxygen or a compound that readily yields oxygen, such as niter, on ignition it undergoes a *redox* reaction of such rapidity as to form an explosion. Gas/air and vapor/air explosions fall into this category. Also, materials such as wood soaked in liquid oxygen can be detonated, which may be regarded as being an aerobic explosion.

Aerobic Explosive – The explosive in which a combustible substance is intimately mixed with oxygen or a compound that readily yields oxygen such as niter, and on ignition undergoes a *redox* reaction of such rapidity as to constitute an explosion. Black powder is an example of aerobic explosive.

Aerogel – Synonym of *Aerosol.*

Aerosol – An aerosol is a heterogeneous two-phase mixture of extremely fine liquid or solid particles, and a gas or vapor, such as smoke, fog. At the time of a volcanic eruption, aerosol is formed naturally.

Synonym: *Aerogel*

Aerosol Explosion – It is a type of *rarified* chemical explosions. *Aerosol* can form explosive mixtures if the liquid or solid particles or the gas is flammable. Low volatile liquids can form extremely fine liquid particles and can suspend in air or gas. For example, heavy hydrocarbons, e.g. heat transfer fluids (HTFs), pump oils, etc., are omnipresent in the manufacturing and process industries, which are capable of forming aerosols when leaked under high pressure. They are generally low volatile and allow the formation of heterogeneous two-phase mixtures that are main constituents for aerosol explosions.

Aerosol explosions may be more devastating than homogeneous vapor-air mixture explosions because of enhanced burning velocities in the heterogeneous mixture and higher enthalpy concentrations in the liquid aerosol phase.

Also see: *Family Tree of Explosions, Heterogeneous Explosion, Homogenous Explosion,* and *Mist Explosion.*

Afire – Blazing, flaming, or on fire.

After-damp – Poisonous mixture of gases formed by the explosion of *firedamp* in a coalmine. The mixture of gases contains carbon monoxide.

Air Bag – A device, incorporated in a vehicle, which is designed and used to reduce the severity of injury that may occur in a collision. Airbag is the latest safety feature incorporated in an automobile to absorb impact, and is designed to act as a supplemental safety device in addition to a seat belt. Nowadays, airbags are mandatory in developed countries for new cars. Air bag system uses an explosive, namely sodium azide (NaN_3).

The first component of the airbag system is a sensor, which detects head-on collisions and immediately triggers the airbag's deployment. As and when the vehicle decelerates very quickly, as in a head-on crash, an electrical circuit is turned on by the sensor, initiating the process of inflating the airbag. As soon as the electrical circuit is turned on, a pellet of sodium azide (NaN_3) is ignited. An almost instantaneous reaction occurs where the sodium combines with the oxygen while the nitrogen atoms regroup into pairs to form large quantities of nitrogen gas. The nitrogen gas fills a nylon or polyamide bag at a velocity of 250 to 400 kilometers (150 to 250 miles) per hour. This process, from the initial impact of the crash to full inflation of the airbags, takes only 30 - 40 milliseconds.

It is an example of the peaceful use of an explosive.

Synonym: Gas Bag.

Air Blast – Air blast is an airborne pressure wave or acoustic transient generated by an explosion. It may be caused by burden movement or by the release of expanding gas into the air. It may or may not be audible.

Air Burst – An above the ground burst of a projectile or bomb.

Airdox – A blasting device that is based on compressed air. The system uses 700 kg/cm^2 (10,000 psi) compressed air to break undercut coal. Airdox does not ignite a gassy or dusty atmosphere.

Also see: *Armstrong Blasting Process.*

Air Loader – A device to charge prilled ANFO blasting agent into borehole when pouring cannot charge the free-running prills. In this process the charge is loaded into a pressure vessel at an air pressure of about 60 psi (4 kg/cm^2). A portable machine that works on the injector principle may also be used.

Air Terminal – A component of a lightning protection device, or a *magazine*, designed to receive direct attachment of the lightning flash and transfer the current to the down conductor.

Ajax Powder – Early *permitted/permissible* dynamite containing ammonium oxalate as a cooling agent. Composed of wood flour, *potassium perchlorate*, *nitroglycerine*, ammonium oxalate, and small quantities of *collodion cotton* and *nitrotoluenes.*

Alex 20 – An explosive consisting of 80% *Composition B* and 20% *aluminum.*

Aliphatic Explosive Compound – Aliphatic organic compounds belong to the alkane (single bond saturated hydrocarbon C−C), alkene (double bond unsaturated hydrocarbons C=C) and alkynes (triple bond unsaturated hydrocarbons C ≡ C). Aliphatic explosives fall into both the open chain and cycloaliphatic groups.

The major sources of oxidizer in most aliphatic explosives are from the nitrate ester group ($-ONO_2$) and the nitramine group ($-NH-NO_2$). *Nitroglycerin*, CH_2 NO_2). $(CH(ONO_2).CH_2(ONO_2)$ and *Pentaerythritol tetranitrate*, $C(CH_2-ONO_2)_4$ are examples of nitrate ester explosives. *RDX*, *Cyclonite*, Cyclo-1,3,5-trimethylene-2,4,6-trinitramine, the empirical formula of which is $(CH-NH-NO_2)_3$ / $(CH_2N.NO_2)_3$, is an example of *nitramine explosive.*

All-fire – All-fire means the minimum energy that has to be applied to a bridge wire circuit for a reliable ignition of the surrounding explosive material under a specified condition.

All-Fire Current – The minimum ampere (or watt) level that has to be applied to a bridge wire circuit for a reliable ignition of the surrounding explosive material without regard to the time of operation. It is recommended that operations at all-fire level be avoided.

Alpha-Trinitrotoluene – Chemical name for *TNT.*

Aluminized Explosive – A high explosive to which aluminum powder or flake has been added to improve the *strength.*

Synonym: *Metallized Explosive.*

Also see: *Energized Explosive.*

Aluminum – Aluminum metal, usually in finely divided particle or flake form, commonly used as a fuel or sensitizing agent in explosives and blasting agents.

Aluminum Flare – Flares used in aviation for signaling and for illuminating landing fields and military objectives. It is composed of a mixture of chlorates, perchlorates, or nitrates, with aluminum powder or flakes, and a binder, such as shellac, and with or without addition of sulfur. It burns with an intensely brilliant light, white or colored, according to composition.

Aluminum Powder – Aluminum powder improves the efficiency of an explosive. The reaction product aluminum oxide is a solid; no gaseous product is formed. The addition of aluminum results in a considerable gain in the heat of explosion, and a higher temperature is imparted to the fumes, because the heat of formation of the aluminum oxide is very high (396 kcal/mol=1658 kJ/mol; 3883 kcal/kg =16260 kJ/kg). The performance effect produced by aluminum powder is frequently utilized in commercial explosives and also in composite propellants.

Widely used mixtures of explosives with aluminum powder include Ammonals, DBX, HBX-1, Hexal, Minex, Minol, Torpex, Trialenes, Tritonal and Hexotonal.

Amatol – A low cost, high explosive made by admixing *trinitrotoluene* (*TNT*) with *ammonium nitrate*. It has commercial and military applications. Amatol is sometimes used as a bursting charge in high-explosive projectiles. It gives greater flame, and is practically smokeless, due to the excess of oxygen provided by the ammonium nitrate. Sometimes used as a *bursting charge* in high-explosive projectiles. The ideal proportions to produce complete combustion are given by the following equation:

$$21NH_4NO_3 + 2C_6H_2(NO_2)_3.\ CH_3 = 14CO_2 + 47H_2O + 24N_2.$$

Amatol Booster – See *Booster, Amatol.*

American – Permissible explosive used in coalmines.

Amid Powder – Synonym of *Amidpulver.*

Amidpulver – A powder consisting of potassium nitrate, ammonium nitrate and charcoal. It gives a flashless and almost smokeless discharge when fired from a gun but has the disadvantage of being highly hygroscopic. German Army in World War I used Amidpulver as cannon powder.

Synonym: Amid Powder.

Ammonal – An explosive of the *ammonium nitrate* class. Consists of *ammonium nitrate*, *trinitrotoluene* (*TNT*), and powdered *aluminum*. It resembles *amatol*, but contains aluminum powder in the admixture. Ammonal is sometimes used as a bursting charge in high-explosive projectiles, and produces bright flashes on detonation.

Ammonia Dynamite – See *Dynamite, Ammonia.*

Synonym: Extra Dynamite.

Ammonia Gelatin – See *Gelatin, Ammonia.*

Ammonia Gelatin Dynamite – See *Dynamite, Ammonia Gelatin*

Ammonium Dynamite – See *Dynamite, Ammonium.*

Ammonium Gelatin Dynamite – See *Dynamite, Ammonium Gelatin.*

Ammonium Nickel Sulfate – See *Nickel Ammonium Sulfate.*

Ammonium Nitrate – See *Nitrate, Ammonium.*

Ammonium Nitrate Explosive – See *Nitrate Explosive, Ammonium.*

Compare: *ANFO.*

Ammonium Nitrate, Prilled – Prilled Ammonium Nitrate: Ammonium nitrate in a pelleted or prilled form.

Ammonium Perchlorate, NH_4ClO_4 – Ammonium perchlorate is a salt that forms colorless or white rhombic and regular crystals. It is soluble in water. It decomposes at 150°C, and the reaction is explosive at higher temperatures. Used as an oxidizer in solid propellant.

Ammonium Perchlorate Explosive – Ammonium Perchlorate, as a constituent of explosives, possesses the advantage that it contains a high proportion of available oxygen and only produces gaseous products. However, the gaseous products include the poisonous gas hydrogen chloride. The formation of hydrogen chloride can be prevented by adding an equivalent quantity of a substance such as sodium nitrate, which will yield a base to combine with the chlorine. Ammonium perchlorate explosives have the following composition:

Ammonium perchlorate - 50%, Dinitro-toluene - 15%, Sodium nitrate - 30% and Castor oil - 5%.

Ammonium Picrate – The compound ammonium picrate, $NH_4C_6H_2O(NO_2)_3$, exists in stable yellow and metastable red forms of orthorhombic crystals. It is the most stable of military explosives. In case an explosive charge has to endure a great amount of physical shock before detonation, ammonium picrate is used in artillery shells and armor penetrating ammunition because physical shock does not set it off easily. Sometimes it is used as a bursting charge in armor-piercing projectiles.

Synonyms: *Explosive D,* and *Ammonium Trinitrophenolate.*

Ammon Gelignite – A secondary high nitroglycerine gelatin explosive containing ammonium nitrate as the prime oxidizing ingredient.

Ammonium Periodate – Ammonium Periodate, NH_4IO_4, has no known use as an explosive. It is used as a chemical reagent. Storage and handling must be done with great care, because a simple abrasion or impact at ordinary temperatures is enough to cause it to detonate violently. It must be kept in a cool, isolated area, away from acute fire hazards, excessive vibration, and shock. The proper method of disposal is to dissolve it in water.

Ammunition – The term Ammunition describes a device that contains explosives, not the explosives themselves.

With respect to a firearms system, ammunition is the consumable component of the system. It is obviously needed to fire a gun.

Ammunition is also a generic term for all kinds of missiles, explosives and pyrotechnic devices. The term "ordnance" includes non-offensive military items.

Ammunition is defined in one Explosives Act as any explosive when the same is enclosed in any case or contrivance, or is otherwise adapted or prepared to form:

- a cartridge or charge for small arms, cannons or other weapons;
- a safety fuse or other fuse for blasting or for shells;
- a tube for firing explosives; or
- a percussion cap, detonator, fog signal, shell, torpedo, war-rocket or any other contrivance.

Ammunition, Blank – Blank ammunition: Ammunition that contains no projectile but contains a charge of low explosive, such as black powder, to produce a noise. Such ammunition is used in training, in signaling and in firing salutes.

Ammunition Data Card – A card prepared for the identification of each individual lot of ammunition manufactured, giving the type and composition of the ammunition. The card also identifies the components of the ammunition by lot number and manufacturer. Instructions for handling the ammunition may also be included in the card.

Ammunition, Drill – Drill Ammunition: Ammunition having no explosive charge as used in practice and training.

Ammunition, Fixed – Fixed Ammunition, usually termed 'cartridge', means ammunition in which the projectile is permanently attached or crimped to a case that contains the primer and the propellant as distinct from separate-loading ammunition.

Ammunition, Illuminating – *See Illuminating Ammunition.*

Also see: *Ship Distress Signal.*

Ammunition, Live – See *Live Ammunition.*

Ammunition, Practice – See *Practice Ammunition.*

Ammunition, Proof – See *Proof Ammunition.*

Ammunition, Separated – See *Separated Ammunition.*

Ammunition, Separate-Loading – See *Separate-Loading Ammunition.*

Ammunition, Small Arms – See *Small-Arms Ammunition.*

Ammunition Primer, Small Arms – See *Small-Arms Ammunition Primer.*

Ammunition, Tracers for – See *Tracers for Ammunition.*

AN – Acronym for *Ammonium Nitrate.*

Anaerobic Explosion – Anaerobic explosion is a type of chemical explosion that occurs in the absence of free oxygen. Under the influence of external forces such as heat or shock, the molecular structure of substance in which the molecule contains carbon and/or hydrogen as well as oxygen but in which the carbon and the hydrogen typically are separated from the oxygen by a nitrogen atom, becomes unstable and an internal *redox* reaction takes place, This reaction liberates nitrogen, water vapor, and oxides of carbon at a high temperature.

Anaerobic Explosive – A chemical explosive substance in which the molecule contains carbon and/or hydrogen as well as oxygen, but in which the carbon and the hydrogen typically are separated from the oxygen by nitrogen atoms. Under the influence of external forces such as heat or shock the molecular structure becomes unstable and an internal *redox* reaction takes place, liberating nitrogen, water vapor, and oxides of carbon at a high temperature. Such an explosive is regarded as an anaerobic explosive. Most military and commercial explosives fall into this category.

Anchor, Shot – See *Shot Anchor.*

ANFO – An acronym for a mixture of Ammonium Nitrate and Fuel Oil. This mixture is used in blasting works in the mines and quarries, where no methane gas, coal dust or water is present.

Mixing ammonium nitrate with a carbonaceous fuel, usually fuel oil, makes up ANFO. The addition of about 6% intermediate flash point fuel oil that, among other things, helps keep it dry. Ammonium nitrate may also be in an emulsion with fuel oil.

An optimum ratio of the mixture is 94.5% AN to 5.5% fuel oil, which gives the maximum energy release and an oxygen-balanced mixture producing non-toxic fumes (carbon dioxide, nitrogen and water vapor). Raising or lowering the AN content respectively produces excess nitrous oxides and carbon monoxide, both of which are poisonous.

The application technique of ANFO blasting agents is very much easier nowadays because the material, which has a strong tendency to agglomeration, is commercially produced as porous prills. Prills are granules solidified from the liquid melt, sufficiently porous to take up about 6% of the fuel oil required to produce oxygen balance.

Porous ammonium nitrate prills mixed with fuel oil, generally loaded un-cartridged (loose) by free running or by means of an *air loader* is extensively used. ANFO is generally mixed onsite by the user.

ANFO is an economical and efficient blasting agent if mixed and used properly. It works well when primed in dry holes. Though it is inexpensive, it requires substantial heat and/or impact to detonate. That is why pentolite or some other explosive is used as a booster to make it detonate.

There are varieties of ANFO. A mixture of ANFO and *Emulsion Slurries,* called heavy ANFO which has higher loading densities than poured ANFO alone.

An addition of Aluminum in the form of powder or flake contributes to a higher heat of explosion due to the formation of Al_2O_3; i.e., producing a higher energy output than a straight ANFO. A mixture of 87.6% AN, 2.5% fuel oil and 9.9% Al is the optimum ratio that gives the maximum heat of explosion.

It is a low density explosive with poor water resistance. Its density is 0.9 g/cm^3. Higher density and water resistant explosives are obtained using slurries and emulsions.

ANFO must be stored in a dry, indoor location at a temperature from -35°C to 35°C.

Also see: *Dynamite, Slurry, Emulsion, Water Gel, Cast Booster, and Ammonium Nitrate Explosive.*

Anhydrous Hydrazine – See *Hydrazine, Anhydrous.*

Anilite – An explosive mixture composed of liquid NO_2 and carbon bisulfide or gasoline. Anilite is very sensitive to shock. When it detonates it acts as a powerful high explosive.

Antipersonnel Bomb – The bomb designed to destroy, maim, or obstruct military personnel.

Applications of Explosives – The major and most spectacular use of explosives has been in warfare. High explosives have found their use in bombs, explosive shells, missile warheads, projectiles and torpedoes. Low explosives, such as gunpowder and the smokeless powders, have been extensively used as propellants for bullets and artillery shells.

The most important peaceful use of explosives is to break rocks in mining. Blast holes are drilled in the rock and filled with any of a variety of explosives; the explosive is then fired. Use of safety explosives in coalmines is a must. The safety explosives produce little or no flame and explode at low temperatures to prevent secondary explosions of mine gases or dust.

Application Time – The time for which the electric current is applied while firing an electric detonator.

APU – Acronym for *Auxiliary Power Unit.*

Ardeer Double Cartridge Test – See *A.D.C.Test.*

Arm – 1. *Weapon.* 2. In blasting industry, 'Arm' is a general term that implies the energizing of electronic and electrical circuitry, which in turn controls power sources or other components used to initiate explosives. The arming operation completes all steps preparatory to electrical initiation of explosives except the actual fire signal.

Arming, Delay – See *Delay Arming.*

Arm, Small – See *Small Arm.*

Armstrong Blasting Process – See *Blasting Process, Armstrong.*

Arm-to-Arm – As applied to fuses, the changing from an armed condition to a state of readiness for initiation.

Aromatic Explosive Compound – A special ring compound, the six-carbon ring with three double bonds, is known by its common name *Benzene.* All organic compounds that contain the benzene ring are included in a class called aromatic compounds. The benzene is the basic building block of aromatic compounds.

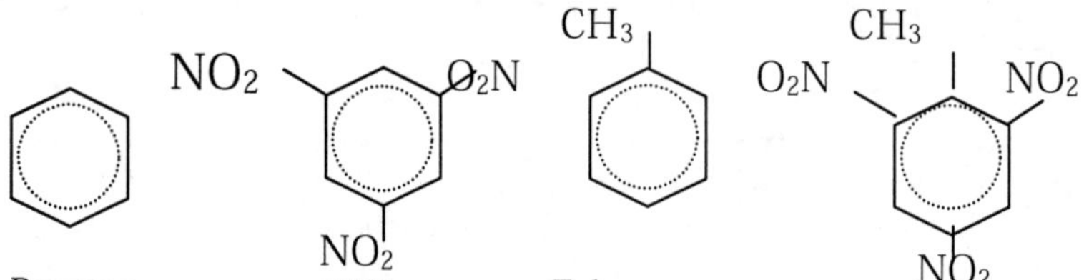

Benzene TNB Toluene TNT

The simplest of the aromatics, and structurally the basis for all of the others in this family, is TNB. TNT and TNX (Trinitroxylene) are examples of mono-substituted and poly-substituted TNBs, respectively.

Arsenic Disulfide – A red, orange, or black compound As_2S_2. It exists as monoclinic crystals, insoluble in water. It is used in fireworks. It occurs naturally as *realgar.*

Arsenic Trioxide – A white glassy mass or powder As_2O_3. It is used in smoke-producing compositions.

Arsenious Oxide – Synonym of *Arsenic Trioxide.*

Artificial Barricade – See *Barricade, Artificial.*

Artillery – Literally, artillery means implements for warfare, weapons for discharging missiles, etc. The term is now applied to heavy firearms. It includes a variety of long-range guns that fire their shells with rapid muzzle-velocity in a low arc; howitzers that fire on a high trajectory at relatively nearby targets; antiaircraft guns that fire rapidly and at high angles; armor-piercing antitank guns; and such other munitions used in support of infantry and other ground operations.

ASA – Mixtures of *Lead Azide*, *Lead Styphnate,* and *Aluminum.*

ATF – Abbreviation for Bureau of Alcohol, Tobacco, Firearms, and Explosives, U.S Department of the Treasury, which enforces explosives control, safety and security.

Atom Bomb – A nuclear weapon whose violent explosive power is due to the sudden release of atomic energy resulting from the splitting or fission of nuclei of a heavy chemical element such as uranium ^{235}U or plutonium ^{239}Pu by neutrons in a very rapid chain reaction. The explosive power of an atom bomb is measured in terms of tens of kilo tons of TNT. The fast, uncontrolled nuclear fission reaction is capable of causing widespread destruction.

The bombs dropped on Hiroshima and Nagasaki (1945) were of this type.

Synonyms: A-bomb, Atomic Bomb and Fission Bomb.

Also see: *Hydrogen Bomb*, *Fusion Bomb*, and *Nuclear Weapon.*

Atomic Bomb – Synonym of *Atom Bomb.*

Atomic Explosion – See *Nuclear Explosion.*

Authorized Explosive – An explosive included in a list published by a statutory authority authorizing it to be manufactured, imported, possessed, and used in the country.

Auto-ignition Point – The auto-ignition point refers to the temperature that a substance must reach before it will ignite in the absence of a flame. Different substances have different auto-ignition temperatures.

Synonyms: Combustion Point, and Self-ignition Point.

Automatic Fire – A ball of quicklime and asphalt that spontaneously ignited on coming into contact with water was discovered and so named by Alexander VI of the Roman Empire during 222-235 AD.

Auxiliary Explosive – The explosives that fall in the intermediate range of sensitivity are termed as auxiliary explosives. In broad classification auxiliary explosives are high explosives. The explosives used as auxiliary explosives are less sensitive than the primary high explosives that are employed in initiators, primers, and detonators. However, they are generally more sensitive than those high explosives used as filler charges or bursting explosives. *Boosters* are auxiliary explosives. Other examples of auxiliary explosives include *Tetrytol, Pentaerythritoltetranitrate (PETN), Trinitrotoluene (TNT), Trinitrophenylmethylnitramine (Tetryl.*

Auxiliary Power Unit – A propellant-powered device used to generate electric or fluid power.

Acronym: *APU.*

Auxplosive Group – In a high explosive, the atom groupings that are commonly present but which by their presence does not produce an explosive. However, they may have an effect upon the power of an explosive.

Axial Priming – A system of *priming* blasting agent. In axial priming, a core of priming material extends through most or all blasting agents.

Azides – Azides are derivatives of hydrazoic acid, HN_3. An azide is a compound of hydrogen or a metal and the mono-valent $^{-}N_3$ radical. The azides as a group are one of the few industrially produced explosives that contain no oxygen. Azides are typically unstable compounds that can violently dissociate (explode). Mercury and lead azides are good examples of the azide type of explosives. Mercury azide is more sensitive than lead azide, even it is more sensitive than mercury fulminate. Like fulminates, azides can be used in detonators and priming compositions as well. Due to its high temperature of ignition, lead azide is not easily ignited by the 'spit' of a safety fuse. However, the difficulty can be overcome by the addition of lead styphnate for instance, which is more readily ignited. It is also used in detonating rivets with the addition of silver acetylide or tetracine to lower its ignition temperature. Azides are used as substitutes for mercury fulminate.

Azides are stored or transported wet. When an azide is packed in bulk it should contain at least 20% water.

Also see: *Barium Azide*, *Lead Azide*, *Mercury Azide,* and *Sodium Azide*.

Back-Blast – The rearward blast of gases from the breech of recoil-less weapons and rockets upon burning of the propellant charge.

Synonym: Breech-Blast.

Backfire – A bang sound made by the explosion of fuel in the manifold or exhaust of an internal combustion engine.

Bag, Air – See *Air Bag.*

Bag, Gas – Gas Bag: Synonym of *Air Bag.*

Ball Cartridge – A *cartridge* that contains a *projectile.*

Ballistic Bomb – Synonym of *Manometric Bomb.*

Synonyms: Bichel bomb, Closed Bomb, Manometric Bomb, and Pressure Bomb.

Ballistic Mortar – An instrument used in laboratories for measuring the relative power or strength of explosive materials.

Ballistite – Alfred Nobel's invention in 1887; made by mixing 40% nitrocellulose and 60% nitroglycerin blended together with diphenylamine. Nitroglycerine increases the energy of nitrocellulose and results in a cleaner burning explosive (smokeless powder). Ballistite was particularly useful as a propelling charge in small arms and mortar ammunition, which could be manufactured relatively safely compared with nitrocellulose propellants. The British developed a number of similar products under the generic name *cordite.*

Ball Powder – Nitrocellulose powder in a uniform spherical form. Nitrocellulose smokeless powder is produced in a uniform spherical form by a simplified process in which stabilization, gelatinization, and solvent recovery are carried out in one operation. It is used as a propellant for ammunition up to 37 mm caliber.

Bang, Big – Big Bang: The cosmic explosion of a single mass of compressed material, which marked the beginning of the universe according to the *big bang theory.*

Also see: *Big Bang Theory.*

Banger – A pyrotechnic that contains a small explosive charge and fuse in a heavy paper casing.

Barium Azide – This compound is a sand-like crystalline solid, whose empirical formula is $Ba(N_3)_2$. Among the salts of hydrazoic acid, Barium Azide is intermediate in its explosive character.

Also see: *Azide.*

Barium Nitrate – Barium Nitrate, $Ba(NO_3)_2$, is a colorless crystal or white powder. It gives green light upon burning. It may be used in pyrotechnic compositions, in blasting explosives, and in some compound propellants.

Barium Peroxide – Barium Peroxide, $Ba0_2$, is a grayish-white powder. It is used in primer combination with aluminum for thermit bombs and in tracer ammunition.

Barrel, Blasting – See *Blasting Barrel.*

Barricade – An approved natural or artificial intervening barrier of such type, size, and construction as to limit in a prescribed manner the effect of an explosion on nearby structures or personnel.

Barricade, Artificial – Artificial Barricade: means an artificial mound or riveted wall of earth, around an explosive storage magazine, of a minimum thickness of one meter (3.3 feet).

Barricaded – The term 'Barricaded' means that a building containing explosives is effectually screened from a magazine, building, railway, or highway. This is either done by a natural barricade or by an artificial barricade of such height that a straight line from the top of any sidewall of the building containing explosives to the eave line of any magazine, or building, or to a point 3.6 meters (12 feet) above the center of a railway or highway, will pass through such intervening natural or artificial barricade.

Barricade, Natural – Natural Barricade: The expression 'natural barricade' means natural features of the ground, such as hills, or timber of sufficient density that the surrounding exposures which require protection cannot be seen from the *magazine* when the trees are bare of leaves.

Barrier, Personnel – See *Personnel Barrier.*

Base Charge – The main charge of high explosive, in the base of a detonator, which is the main charge of the detonator and is fired by the *priming charge*. The base charge makes the major contribution to the power of the device.

Bay – A room, cubicle, cell or work area wherein a single type of explosives activity is performed, which affords the required safety provisions specified for appropriate hazard classification of the activity involved.

BB Cap – A .22 caliber metallic cartridge about 1 centimeter (0.4 inch) in diameter that consists of a rim fire case and a small (about 20 grain) round-nose lead bullet, and in which originally the primer originally served as the propellant.

BB Gun – A smooth bore air gun actuated by a spring-loaded plunger that upon release from the cocked position compresses the air behind the pellet and propels it from the tube.

B-black Powder – The sodium nitrate version of *black powder.*

B-blasting Powder – The sodium nitrate version of *blasting powder.*

Behavior of Explosive on Ignition – Upon ignition an explosive can *burn* at a controlled rate if heat and gas are free to escape. It can *deflagrate* if the burning rate increases due to thermal conduction and radiation. It can *detonate* if the deflagration front reaches shock wave conditions, i.e. heat and pressure reinforce the shock front; rise of temperatures and pressures to about 3500°C and 300000 kgs/cm^2 (4300000 psi) respectively.

Bengal Fire – A variety of colored fires or flares. Usually a bluish white light, used formerly in signalling and for illumination in theatres. This material is composed of a deflagrating mixture of potassium nitrate or saltpeter, sulfur, and arsenic disulfide (realgar).

Synonyms: *Bengal light*, and *Indian fire.*

Bengal Light – See *Bengal Fire.*

Benzine Torch – Synonym of *Molotov Cocktail.*

Bichel Bomb – Synonym of *Manometric Bomb.*

Bickford Fuse – Commercial name for one brand of slow burning *fuse.*

Big Bang – See *Bang, Big.*

Big Bang Theory – A theory in astronomy that the universe originated billions of years ago from the explosion of a single mass of dense matter.

Binary Explosive – An explosive formed by mixing or combining two non-explosive materials, such as *ammonium nitrate* and *nitro-methane.* Individual component of a binary explosive is generally classified as non-explosive. Each component is stored separately and mixed at the job site to form a detonable high explosive. Binary explosives are blasting explosives.

This type of explosive is commonly used for intermittent or limited usage due to mixing and time requirements.

Synonym: *Two-component Explosive.*

Biodegradable – The term 'biodegradable' means the property of a material to decompose naturally.

Black Blasting Powder – It is a grained *black powder,* which is normally supplied as a glazed material. A small amount of graphite is used to impart the glaze finish to the final product.

Black Match – See *Match, Black.*

Black Powder – It is an intimate mechanical mixture of saltpeter (potassium nitrate), sulfur and charcoal, which is mostly pressed, granulated and classified into definite grain fractions. The standard composition of black powder is: 75% potassium nitrate (saltpeter), 10% sulfur and 15% charcoal. Corresponding compositions based on sodium nitrate are known as B-*Black Powder.* Sodium nitrate version of black powder, sometimes called *soda powder*, is more powerful than original black powder (saltpeter version).

Black powder is a *deflagrating low* explosive. It is sensitive to impact, friction and sparks.
It is the oldest known explosive. The exact history is obscure. *Saltpeter*, the basic ingredient of black powder, is thought used historically as early

as the 10th century. Black powder continued to be the main explosive material until the late-1800s. It is one of the most important substances ever known to mankind because its manufacture brought forth modern chemistry and modern science. Its use revolutionized warfare and ultimately played a large part in the alteration of European patterns of living up until modern times. Gunpowder was the only explosive in wide use until the middle of the 19th century, when nitroglycerine-based explosives superseded it.

Black powder was called *gunpowder* until nitrocellulose powder became the most widely used type. After that, *nitrocellulose powder* was called *gunpowder* and black powder gained its own identity.

Because of its low energy, poor fume quality, and extreme sensitivity to sparks, the use of black powder has decreased. It has been largely replaced as a propellant by the so-called smokeless powder. Today, it is used primarily in the powder of safety fuse and in pyrotechnics. It is also used for initiating charges, primers and fuses in military ammunition. It is suitable for controlled blasting in which the treatment of the stone must be mild.

Synonyms: Black Blasting Powder, Gunpowder, A-Blasting Powder, and B-Blasting Powder.

Also see: *Gunpowder, A-Black Powder,* and *B-Black Powder.*

Blank Ammunition – See *Ammunition, Blank.*

Blank Cartridge – A cartridge that does not have a projectile.

Blank Cartridge for Weapon – It consists of a cartridge case with primer and propellant or black powder but no projectile. Blank cartridge for weapon is used for training, saluting, propelling charge, etc.

Blank-fire Powder – Synonym of *EC Smokeless Powder.*

Blank Fire, EC – EC Blank Fire: Synonym of *EC Smokeless Powder.*

Blast – 1. The detonation of explosives to break rock; 2. To employ an explosive to shatter or expose something; 3. Sudden air pressure created by the discharge of a gun or the explosion of a charge.

Blast, Air – See *Air Blast.*

Blast Area – The area of a blast within which the influence of flying rock missiles, gases, and concussion extends.

Blast, Back – See *Back-Blast.*

Blast Bomb – Blast bomb is a type of improvised explosive device (IED). It is usually a form of homemade grenade, which is thrown at a target. Ordinary fireworks are sometimes used in a similar fashion to blast bombs.The term is used in Northern Ireland.

Blast, Breech – Breech-Blast: Synonym of *Back-Blast.*

Blaster – A person qualified through knowledge, training, or experience to be in charge of and responsible for the loading and firing (or detonating) explosives in a blasting operation. A blaster also has to have qualifications relating to transporting, storing, handling, and use of explosives, and have a working knowledge of statutory laws and regulations that pertain to explosives.

Synonyms: Shot firer, powder man, and powder monkey.

Blast Hole – Synonym of *Bore Hole.*

Blast Hole Charger – A portable set of devices consisting of a prilled explosive reserve tank feeding into an air-activated loading tube. The blast hole charger permits rapid loading of prilled explosives into blast holes drilled in any direction.

Blast Hole Drill – A drilling machine used to produce holes in which an explosive charge is placed.

Synonym: Shot Hole Drill.

Blastine – A *chlorate mixture* type explosive.

Blasting – The operation of breaking up of heavy masses (as of rock or other material), moving material, or generating seismic waves by boring a hole, filling with explosive charge, and firing.

After ammunition, the second most significant usage of commercial explosives is for blasting. The activity must only be carried out by authorized operators with due precautions.

Blasting Accessories – Any non-explosive devices and materials used in blasting.

Blasting Agent – A non-detonator sensitive explosive that must be initiated by a *booster*, or that requires a *primer* to detonate. It is an explosive material under UN Classification 1.5 D, which is insensitive to initiation by a #8 detonator.

The most common blasting agent is *ANFO*, but many gels, slurries, and emulsions are also blasting agents. Blasting agent categories are a classification for the storage and transportation of some explosives since they are less sensitive to initiation and, therefore, can be stored and transported under different regulations than would be used for more sensitive high explosives.

Synonym: Detonator Insensitive Explosive.

Blasting Agent, Dry – Dry Blasting Agent: The term "dry blasting agent" describes any material in which no water is used in the formula. The early dry blasting agents employed fuels of solid forms of carbon or coal dust combined with ammonium nitrate in various forms. The early agents proved less successful since the solid fuels segregated during transportation and provided lower blasting results. It was found that diesel oil mixed with porous ammonium nitrate prills gave the best overall blasting result, and therefore the term ANFO has become synonymous with dry blasting agents.

Blasting Cap – A small tube filled with detonating substances used to detonate high explosives. It is a synonym for detonator. A *detonator* is usually called a blasting cap in the USA. The term 'blasting cap' is generally not used today except to describe the device used in conjunction with safety fuse.

After the invention of *practical detonator,* Alfred Nobel in 1865 invented an improved detonator called a blasting cap, which consisted of a small metal cap containing a charge of mercury fulminate that can be exploded by either shock or moderate heat. The invention of the blasting cap inaugurated the modern use of high explosives.

Also see: *Cap, Detonating Device, Detonator, Fulminate of Mercury, Fulminating Mercury,* and *Mercury Fulminate.*

Blasting Cap Assemblies – Plain detonators assembled with and activated by means of detonating cord, safety fuse or shock tube. The basic detonators may be of instantaneous type or may incorporate delay elements or detonating delays.

Synonym: *Detonator Assemblies, for blasting.*

Blasting Cap, Electric – Electric Blasting Cap: Synonym of *Electric Detonator.*

Blasting Cap, Electric Delay – Electric Delay Blasting Cap: Synonym of *Electric Delay Detonator.*

Blasting Cap, Ordinary – See *Ordinary Blasting Cap.*

Blasting Cartridge – A cartridge containing an explosive charge to be used in a blasting operation.

Blasting Circuit – A shot-firing cord together with connecting wires and electric blasting caps used for the purpose of firing of a blast.

Blasting Circuit, Electric – Electric Blasting Circuit: An electric circuit containing electric detonators and associated wiring.

Blasting Circuit, Series – See *Series Blasting Circuit.*

Blasting Circuit, Series in Parallel – See *Series in Parallel Blasting Circuit.*

Blasting Compound – An explosive substance used in mining and quarrying.

Blasting Cord – See: *Shot-firing blasting cord.*

Blasting, Coyote – See *Coyote Blasting.*

Synonym: Coyote Shooting.

Blasting, Delay – Delay Blasting: The practice of initiating individual explosive charges to detonate at predetermined time intervals by the use of *delay detonators* or connectors, instead of instantaneous blasting where all holes are fired essentially simultaneously.

Antonym: *Instantaneous Blasting.*

Blasting, Detonator for – See *Detonator for Blasting.*

Blasting Dynamite – See *Dynamite, Blasting.*

Blasting, Electric – Electric Blasting: The term 'electric blasting' means the firing of one or more charges electrically by the use of electric blasting caps, electric squibs, or other electric igniting or exploding devices.

Blasting Fuse – Compound, usually gunpowder, designed to burn at a regulated speed when enclosed in a flexible fabric tube, used to ignite a detonator or explode a blasting charge. Types include 'safety' (slow or instantaneous) and detonating.

Blasting Galvanometer – A simple instrument to test electric blasting circuits that enables the blaster to locate breaks, short circuits, or faulty connections before an attempt is made to fire the shot. Misfires may be averted to a great extent with the use of a blasting galvanometer.

Blasting Gelatin – Blasting gelatin is the most powerful industrial explosive and the strongest of all nitroglycerine explosives. Alfred Nobel invented this dynamite composition. It is a colloid produced by gelatinizing nitroglycerine with about 7-8% nitrocellulose. *Nitroglycerine* increases the energy of *nitrocellulose* and results in a cleaner burning explosive. This jelly-like material is translucent, rubber like elastic and unaffected by water.

Varieties of blasting gelatin contain suitable proportions of potassium nitrate and wood-meal.

It has an extremely high velocity and is rated at 100% strength, and taken as a standard of explosive power. Rate of detonation is 7800 meters (25500 feet)/second. It is used for special cases of tunnel-driving, shaft-sinking, deep-well shooting, and submarine work.

Blasting gelatin is used as a comparative explosive in determinations of relative weight strength. (See *Ballistic Mortar.*)

Synonyms: Nitroglycerine Gelatin, and Torpedo Explosive No. 1.

Also see: *Nitroglycerine* and *Cellulose Nitrate.*

Blasting, Instantaneous – Instantaneous blasting: The blasting in which all explosive decks, boreholes, or rows of boreholes are fired essentially simultaneously.

Antonym: *Blasting, Delay.*

Blasting Log – A record of information about a specific blast.

Blasting Machine – A portable electrical or electromechanical device (dynamo) that generates enough electric current to detonate electric blasting caps when the machine is actuated.

Synonym: Exploder.

Blasting Machine, Capacitor-discharge – Capacitor-discharge Blasting Machine: A blasting machine in which electrical energy, stored on a capacitor, is discharged into a blasting circuit containing electric detonators.

Blasting Machine, Sequential – See *Sequential Blasting Machine.*

Blasting Mat – A mat of woven steel wire, rope, scrap tires, or other suitable material or construction to cover blast holes for the purpose of preventing flying rock missiles.

Blasting Needle – A pointed instrument for piercing the wad or tamp of a charge of explosive, to permit introducing a blasting fuse.

Blasting Oil – An industrial term, now obsolete, for *nitroglycerine.*

Also see: *Explosive Oil.*

Blasting Powder – A type of *Black powder* that is manufactured in grains or pellets. Used especially for blasting of material, such as coalmines.

Synonym: Black Blasting Powder.

Blasting Process, Armstrong – Armstrong Blasting Process: An extraction method in the USA, especially in coal mining. The highly compressed (700-800kg/cm^2; 10000-12000psi) air in the borehole is suddenly released by means of a blasting tube equipped with a bursting disc. The compressed air is generated underground by a special compressor.

Also see: *Airdox.*

Blasting, Propagation – See *Propagation Blasting.*

Blasting, Secondary – Secondary Blasting: The term 'secondary blasting' is used to mean the reduction of oversized material by the use of explosives to the dimension required for handling, including mud-capping and block holing.

Blasting, Short Delay – See *Short Delay Blasting*.

Blasting Switch – The switch that is used to connect a power source to a *blasting circuit.*

Blast, Muzzle – Muzzle Blast: The sudden air pressure developed in the close proximity of the muzzle of a weapon by the rush of hot gases and air upon firing.

Blast, Primary – See *Primary Blast.*

Blast Shield – A specialty type of portable protective shield that is designed to protect the user from thermal effects, fragments and overpressure. Both bomb technicians and tactical personnel use it.

Blast Site – The area where explosive material is handled during loading, including the perimeter of blast holes and for a distance of 15 meters (50 feet) in all directions from loaded holes or holes to be loaded. In underground mines, 4.5 meters (15 feet) of solid rib or pillar can be substituted for the 15 meters (50 feet) distance.

Blast Tube – A device that is used for the study of shock waves and for calibration of air-blast gauges.

Synonym: Shock Tube.

Blast Wave – See *Wave, Blast.*

Compare: S*hock Wave.*

Blending – Blending, in relation to binary explosives or blasting agents, means the mixing of solid materials by gravity flow, usually induced by the rotation of the vessel.

Blends – Blends mean the mixture consisting of water based explosives or oxidizer matrix and ammonium nitrate or ANFO.

BLEVE – Acronym for *Boiling Liquid Expanding Vapor Explosion.*

Block Demolition Charge – A charge, composed of a high explosive such as TNT, Tetrytol, Composition-C Series and / or Ammonium Nitrate, used for the purpose of general demolition operations, such as tree cutting, breaching and cratering.

Block Hole – A hole drilled into a boulder with the intention of allowing the placement of a small charge of explosive for breaking the boulder.

Boiler – The part of the steam generator where water is converted to steam; usually comprised of metal shells, headers, and tubes which form the container for the steam and water under pressure. Boilers may be treated as a type of pressure vessel.

Boiler Compound – Any chemical substance added to boiler feed water to prevent or inhibit corrosion, foaming, or the formation of scale.

Boiler Efficiency – Ratio of the heat furnished by a boiler in heating and evaporating water to the heat furnished to the boiler by the fuel.

Boiler Feed Water – Water used by boiler to form steam for heat or power.

Boiler Iron – See *Boiler Plate.*

Boiler Plate – Soft low-carbon steel plate for boiler manufacture.

Boiler-plug Alloys – Alloys of melting point below that of tin containing bismuth, lead, tin, cadmium and mercury.

Boiler Priming – Boiling so violently that water in form of spray is carried from the boiler before it can be separated from the steam.

Boiler Scale – Incrustation in boilers due to mineral content of water.

Boiling – Conversion of liquid into vapor with formation of bubbles.

Boiling Liquid Expanding Vapor Explosion – A specific sequence of events commencing with the sudden rupture due to fire impingement of a vessel/system under pressure containing liquefied flammable gas. The release of energy from the pressure burst and the flashing of the liquid to vapor creates a localized blast wave, which is not due to the flammability of the material. However, immediate ignition of the expanding fuel-air mixture leads to intense combustion, creating a fireball that rises from the ground due to buoyancy.

Acronym: BLEVE.

Boiling Point – The temperature at which the vapor pressure of a substance is equal to atmospheric pressure.

Bomb – 1. A bomb is an explosive device, usually some kind of container filled with explosive charge, designed to cause random destruction when detonated. It generates and releases energy very rapidly as a violent, destructive shockwave. In a bomb the projectile or other device carrying an explosive charge is fused to initiate under specific conditions, such as through a timing appliance or upon impact. It is set into position at a given point or is dropped (as from an aircraft), or hurled (as by a mortar) with varying effects depending upon the type used.

The word 'bomb' comes from the greek word *bombos*, which has the same meaning as 'boom' in English. It is not usually applied to industrial explosive devices.

The explosion of a bomb may be triggered by a detonator, fuse, a clock, a remote control, or some kind of sensor, usually pressure (altitude), radar, or contact. An explosive bomb has three main components: a power source, an initiator, and an explosive substance. Other parts, such as timers, switches, relays, and casings, are used simply to control the time and the place of the explosion.

Bombs may be classified into three categories: *conventional*, *dispersive*, or *nuclear*. Based on the method of manufacture, bombs may be categorised as *military bomb* and *terrorist bomb*.

2. A pressure vessel, such as a steel cylinder, for carrying out chemical experiments.

Bomb, A – A-Bomb: Synonym of *Atom Bomb.*

Bomb, Antipersonnel – See *Antipersonnel Bomb.*

Bomb, Atom – S*ee Atom Bomb.*

Bomb, Atomic – Atomic Bomb: Synonym of *Atom bomb.*

Bomb, Ballistic – See *Ballistic Bomb.*

Bomb, Bichel – See *Bichel Bomb.*

Bomb, Blast – See *Blast Bomb.*

Bomb, Butterfly – See *Butterfly Bomb.*

Bomb Calorimeter – Bomb calorimeter is an instrument used for measuring heat of reaction, especially heat of combustion, such as for determining the calorific value of a fuel. It is usually a robust cylindrical metal vessel.

Bomb Calorimeter, Adiabatic – Adiabatic Bomb Calorimeter: Synonym of *Bomb Calorimeter.*

Bomb, Chemical Fuse – See *Chemical Fuse Bomb.*

Bomb, Cherry – See *Cherry Bomb.*

Bomb, Closed – See *Closed Bomb.*

Also see: *Ballistic Bomb.*

Bomb, Cluster – See *Cluster Bomb.*

Bomb, Conventional – See *Conventional Bomb.*

Also see: Bomb.

Bomb, Crawford – See *Crawford Bomb.*

Bomb, Criminal – See *Criminal Bomb.*

Bomb, Dirty – See *Dirty Bomb.*

Bomb, Dispersive – See *Dispersive Bomb.*

Bomb Disposal – It is the process by which hazardous devices are rendered safe. The term is used to describe civilian (Unexploded Ordnance, UXO), military (Explosive Ordnance Disposal, EOD), and public safety (Public Safety Bomb Disposal, PSBT) operations.

Bomb Disrupter – A device that projects a fluid projectile or solid slug into an explosive package such as an I.E.D. or bomb. The intent is to decrease the density of the explosive by causing it to be spread apart from the impact of the projectile - at a speed faster than the reaction time of the initiating system.

Bomb, Dispersive – See *Dispersive Bomb.*

Bomb Drop Test – A test to determine the sensitivity of military explosives as bomb fillers. Bomb drops are made using bombs assembled in the conventional manner, as for service usage, but containing either inert or simulated fuses. The target is usually reinforced concrete.

Bomb, Explosive – See *Explosive Bomb.*

Bomb, Fire – Fire Bomb: Synonym of *Incendiary Bomb.*

Bomb, Fission – See *Fission Bomb.*

Bomb, Fusion – See *Fusion Bomb.*

Bomb Fuse – A bomb *fuse* is a mechanical or electrical device, through which bomb detonation is controlled to cause the detonation at the proper time after certain conditions are met.

A bomb is usually manufactured to withstand reasonable heat and be insensitive to the shock of ordinary handling; hence, the requirements of a bomb fuse. The bomb fuse has the sensitive explosive elements (the primer and detonator) and the necessary mechanical /electrical action to detonate the main burster charge. A mechanical action or an electrical impulse, which causes the detonator to explode, fires the primer. The primer-detonator explosion is relayed to the main charge by a booster charge. This completes the explosive train.

Bomb, Homemade – Homemade Bomb: See *IED.*

Bomb, Hydrogen – See *Hydrogen Bomb.*

Bomb, Illuminating – Illuminating Bomb: Synonym of *Illuminating Ammunition.*

Bomb, Incendiary – See *Incendiary Bomb.*

Bomb, Kiloton – See *Kiloton Bomb.*

Bomb Lance – A lance or harpoon with an explosive head, used in the fishing of whale.

Bomb-let – A small bomb.

Bomb, Letter – See *Letter Bomb.*

Bomb, Mail – Mail Bomb: See *Letter Bomb*, the synonym of mail bomb.

Bomb, Manometric – See *Manometric Bomb.*

Bomb, Military – See *Military Bomb.*

Bomb, Molotov – Molotov Bomb: Synonym of *Molotov Cocktail.*

Bomb, Nail – See *Nail Bomb.*

Bomb, Napalm – See *Napalm Bomb.*

Bomb, Neutron – Neutron Bomb: See *Nuclear Weapon.*

Bomb, Nuclear – Nuclear Bomb: See *Nuclear Weapon.*

Bomb, Parcel – Parcel Bomb: Synonym of *Letter Bomb.*

Bomb, Petrol – Petrol Bomb: Synonym of *Molotov Cocktail.*

Bomb, Photo-flash – See *Photo-flash Bomb*.

Bomb, Pipe – *See Pipe Bomb.*

Bomb, Pressure – See *Pressure Bomb*.

Bombproof – A shelter so constructed or placed as to be secure against the explosive force of bombs or shells.

Bomb, Smoke – See *Smoke Bomb.*

Bomb, Sofar – See *Sofar Bomb.*

Bomb, Splinter – See *Splinter Bomb.*

Bomb Suit – A specialty type of protective suit E.O.D personnel and bomb technicians wear when handling hazardous devices. It is designed and manufactured to provide maximum protection to E.O.D personnel and bomb technicians against thermal effects, fragments, overpressure and impact, which may result if any hazardous device is initiated.

Bomb, Terrorist – See *Terrorist Bomb.*

Synonym: Improvised Explosive Device (IED)

Compare: *Military Bomb.*

Bomb, Thermonuclear – See *Thermonuclear Bomb.*

Bomb, Thermos – See *Thermos Bomb.*

Bomb, Time – See *Time Bomb.*

Bomb-tube – Thick-walled hard glass tube, sealed at one end, about 35-45 centimeters (14-18 inches) long.

Bomb, Vacuum – See *Vacuum Bomb.*

Bomb, Water – See *Water Bomb.*

Bonding, Electrical – See *Electrical Bonding.*

Bonding, Explosion – Explosion Bonding: Synonym of Explosion Welding. See *Welding, Explosion.*

Bonded – The joining of metallic parts to form an electrically conductive path that will ensure electrical continuity and the capacity to conduct safely any current likely to be imposed.

Bonding – An electrical connection between a metal object and a component of the Lightning Protection System (LPS). This produces electrical continuity between the LPS and the object and minimizes electro-magnetic potential differences. Bonding is done to prevent side flash.

Booby Trap – Concealed explosive devices generally attached to some harmless-looking objects. It may be defined as any bomb disguised as something other than a bomb.

Boomer – 1. Specific to the Boomer Shoot – a reactive target made of high explosives. 2. Specific to the gun community – a very high power rifle that makes a louder than normal noise. Magnum rifles may be referred to as boomers.

Boom Powder – See *Powder, Boom.*

Boom, Sonic – See *Sonic Boom.*

Booster – An explosive charge, usually of high detonation velocity and detonation pressure, designed to be used in the explosive initiation sequence between an initiator or primer and the main charge. [Definition courtesy of *ISEE Blasters' Handbook, 17th Edition,* and *IME's SLP 12.*]

Also see: *Cast Booster,* and *Primer.*

Booster, Amatol – Amatol Booster: *Pentolite booster* or *Composition B* booster that contains amounts of ammonium nitrate is called amatol booster.

Booster, Cast – Cast Booster: A solid cap-sensitive high explosive that typically contains TNT as the casting material. Different molecular explosives, such as Pentolite, are added to the melted TNT to increase energy and add sensitivity to the booster. A cast booster attains the property of high detonation pressure, and has excellent water resistance.

Booster, Composition B – Composition B Booster: A booster that contains the *Composition B* with wax added to the mixture. Many of the boosters that are generically given this name are diluted with additional amounts of TNT.

Booster, Extruded – Extruded Booster: Synonym of *Cast Booster.*

Booster, Pressed – Pressed Booster: Synonym of *Cast Booster.*

Booster Charge – The final high explosive component of an explosive train, which amplifies the detonation from the lead or detonator so as to reliably detonate the main high explosive charge. Also used loosely to indicate a reinforcing or augmenting charge.

Booster Explosive – A booster explosive is that component of the explosive train, which functions to transmit and augment the force and flame from the initiating explosive. It ensures the reliable detonation or burning of the main burster charge or propellant charge. The high-explosive booster uses Tetryl or Composition A-5, while the propelling charge uses a black powder booster.

Booster, Pentolite – See *Pentolite Booster.*

Booster, Sodatol – See *Sodatol Booster.*

Booster, Torpex – See *Torpex Booster.*

Bootleg – The part of a drilled borehole that remains relatively intact when the force of the explosion does not break the rock completely to the bottom of the hole. There is a strong likelihood of containing unfired explosive in a bootleg, which should be dealt with accordingly.

Bore Hole – The hole drilled in the material to be blasted, for the purpose of containing the explosive charge.

Synonym: Blast Hole.

Boss – An outlet provided in the generator case for hot gas flow, igniter, pressure measurement, and safety diaphragm.

Branch Line – A length of detonating cord.

Break Gallery – Break gallery is an experimental installation to assess the hazards of firing explosives in coalmines in the presence of breaks.

Breech – Reloadable pressure vessel used to contain a propellant cartridge.

Breech-Blast – Breech-blast is synonym of *Back-Blast.*

Bridge Wire – The very fine filament wire connecting the ends of the leg wires inside an electric blasting cap (detonator), which is imbedded in the ignition charge of the electric blasting cap (detonator). An electric current passing through the wire causes a sudden heat rise, causing the ignition charge to be ignited.

Bridge Wire Detonator – The detonator that is built up by an incandescent bridge made of the resistance wire. The wire, which is made to glow by application of an electric pulse, is covered by immersion in a solution of a pyrotechnical material. This type of detonator is usually used in industrial sector. Copper casing instead of conventional aluminum casing is used in coal mining due to the presence of methane gas.

Brimstone – Synonym of *Sulphur.*

Brisance – It is the shattering power effect of an explosive. Brisance is dependent upon the detonation rate and loading density and the heat of explosion and gas yield.

Detonation rate increases with the increase of the density of the explosive, and the shock wave pressure in the detonation front varies with the square of the detonation rate. To have higher brisance the loading density should be as high as possible. This is particularly true for shaped charges.

Brisant – The term, as applied to explosions, indicates a powerful impulse of short duration.

Briska Detonator – See *Detonator, Briska.*

Bubble Energy – The term 'bubble energy' means the expanding gas energy of an explosive, as measured in an underwater test.

Bulk Density – The mass per unit volume of a bulk explosive such as ANFO. The term is used in connection with packaging, storage or transportation.

Bulk Mix – A mass of explosive material prepared for use on site without packaging, e.g. ANFO.

Bulk Mix Delivery Equipment – Equipment that transports explosive materials in bulk form for mixing or loading directly into blast holes, or both. This equipment is usually a motor vehicle with or without a mechanical delivery device.

Bulk Strength – Bulk strength is the strength per unit volume of an explosive or blasting agent, usually expressed as a percentage of the strength per unit volume of *blasting gelatin.*

Bulk Strength, Absolute – See *Absolute Bulk Strength.*

Bulk Strength, Relative – See *Relative Bulk Strength.*

Bulldoze – See *Adobe Charge.*

Bullet – The only component of a cartridge, which is a single piece of ammunition. This small, cylindrical missile is made of lead, steel, or lead with a steel casing. It is designed to be fired from a small firearm, e.g. musket, revolver, rifle. It is broadly called cartridge.

Bullet, Percussion – Percussion Bullet: A bullet containing a substance that is exploded by being struck inside a firearm.

Bullet-resistant – Bullet-resistant, in relation to explosive magazines, means walls or doors of construction resistant to penetration of a bullet of 150-grain M2 ball ammunition, having a nominal muzzle velocity of 825 meters (2700 feet) per second, fired form a 30 caliber rifle from a distance of 30 meters (100 feet) perpendicular to the wall or door.

When a ceiling or roof of a magazine is required to be bullet-resistant, the ceiling or roof should be constructed of materials comparable to the sidewalls, or of other materials that can withstand penetration of the bullet described above when fired at an angle of 45 degrees from the perpendicular.

In order to determine bullet resistance, the test must be conducted on test panels of empty magazines, which shall resist penetration of five out of five shots placed independently of each other in an area at least one meter by one meter (3 feet by 3 feet).

Bullet-Sensitive Explosive Material – Explosive materials that can be detonated by 150-grain M2 ball ammunition having a nominal muzzle velocity of 825 meters (2700 feet) per second, when the bullet is fired from a .30 caliber rifle at a distance of not more than 30 meters (100 feet) and the test material, at a temperature of 21 to 24°C (70 to 75° F), is placed against a backing material of 1.25 centimeter (1/2-inch) steel plate.

Bureau of Alcohol, Tobacco, Firearms and Explosives – See *ATF.*

Bureau of Explosives – See *AAR's Bureau of Explosives.*

Burning – The term 'burning' denotes any oxidation reaction including those that introduce atmospheric oxygen.

Burning, Cigar – See *Cigar Burning.*

Burning, End – End Burning: The term 'end burning' is used to describe a solid propellant grain, which is inhibited so as to burn from one end only, so that burning progresses in the direction of the longitudinal axis.

Burning Mixture – Synonym of *Low Explosive* or *Deflagrating Explosive.*

Burning, Progressive – See *Progressive Burning.*

Burning Rate – The linear rate of evaporation of material from a liquid pool during a fire, or the mass rate of combustion of a gas or solid. The context in which the term is used should be specified.

Burning Train – Step-by-step arrangement of charges in pyrotechnic or propellant by which the initial fire from the primer is transmitted and intensified until it reaches and sets off the main charge. An explosive bomb, projectile, etc., uses a similar series.

Synonyms: Explosive Train, and Igniter Train.

Burst – Explosion of a projectile in the air, or when it strikes the ground or target.

Burst, Air – See *Air Burst.*

Burster – Explosive charge used to break open and spread the contents of chemical projectiles, bombs or mines. The burster is no larger than what is required to burst the case and disseminate the contents.

Bursting Charge – A quantity of an explosive designed to produce effect by blast or fragmentation.

Bursting-Charge Explosive – The bursting-charge explosive is characterized by being relatively insensitive and having high brisance or shattering power.

Also see: Ammonium picrate, High Explosives, Picric Acid, Trinitrotoluene.

Bursting Explosive – Synonym of *Disrupting Explosive.*

Burst, Pressure – See *Pressure Burst.*

Burst, Separating – Separating Burst: Method of ejecting the contents of a projectile by means of a charge of propellant that breaks the projectile into two approximately equal parts, along a specially designed circumferential shear joint.

Bursting Time – The time between the application of the electric current and the explosion of the detonator while firing an electric detonator.

Bus Wire – See *Wire, Bus*

Butterfly Bomb – The first *cluster bomb* operationally used during World War II to attack both civilian and military targets, commonly known as the Butterfly Bomb. The bomb was so named because the thin cylindrical metal outer shell hinged open when it was dropped and resembled a large butterfly.

Also see: *Cluster Bomb.*

C-4 – A composite military explosive, the rate of detonation being 26,400 feet per second. It is a *plastic bonded explosive*, (*PBX*) containing approximately 91% RDX and 9% non-explosive plasticizer (plastic stabilizer: motor oil - 1.6%, polyisobutene - 2.1%, di-2-ethylhexy, sebacate - 5.3%). It is moldable, pliant and relatively stable and insensitive to impact and friction. White to light brown in color. C4 is effective in temperatures between -70 and +170°F, but loses its plasticity in colder temperatures.

Synonym: *Composition C4.*

Cake, Mill – See *Mill Cake.*

Calcium Carbide – Chemical combination of carbon and Calcium. Chemical names are Calcium carbide and Calcium dicarbide. Commonly it is called Carbide. It is also known as carbide of commerce. Chemical formula is CaC_2. It reacts with water to evolve *acetylene* gas (ethyne).

Calcium carbide is a noncombustible solid. Upon contact with water, carbide readily evolves flammable acetylene that is responsible for the so-called calcium carbide fires.

Calcium carbide and water or water vapor will react in any environment, including oxygen-deficient and inert atmospheres, to generate acetylene.

Calcium carbide is a grey, brown, or black granular solid that reacts with water to form acetylene and carbide lime. When crushed, calcium carbide is a rock-like solid with sharp, angular surfaces representing irregular fracture planes. The older the calcium carbide, the smoother these surfaces become due to reaction with moisture in the air. Calcium carbide has a unique odor that has been described as garlic-like, due primarily to trace impurities given off when the calcium carbide reacts with atmospheric moisture.

Calcium carbide itself will not burn or explode. It is made in an electric furnace and is tapped from the furnace white hot at 3600°F (1980°C) into molds open to the air.

It reacts readily with water, water moistened materials, or moisture in any form (fog, mist, spray or vapor) to form acetylene and calcium oxide.

Water (in any form) shall be kept away from contact with calcium carbide. One of the more dangerous conditions is water contacting the bottom of a pile of calcium carbide. The heat of reaction of calcium carbide and water generates warm, moist acetylene that reacts with the upper layers of carbide forming more acetylene and higher temperatures. Continued access of water to the bottom of the pile will continue heating the carbide until the carbide temperature exceeds the temperature necessary to ignite an acetylene-air mixture.

When moist air contacts calcium carbide dust, the acetylene-air mixture generated can combust spontaneously due to the heat generated when the moisture reacts with calcium carbide.

Synonyms: *Calcium Dicarbid, Carbide*, and *Carbide of Commerce.*

Calcium Dicarbide – Synonym of *Calcium Carbide.*

Caliber – Diameter of a projectile or that of the bore of a gun.

Calorimeter – A device for the measurement of heat of combustion in a compressed oxygen atmosphere; or heat of explosion in an inert gas (such as argon) atmosphere. The inert gas atmosphere is used for propellants, pyrochemical mixtures and explosives that react without outside oxygen.

Also see: *Bomb Calorimeter.*

Calorimeter, Gas – See *Gas Calorimeter.*

Candle – See *Roman Candle.*

Candle, Roman – See *Roman Candle.*

Candlepower – It is the luminous intensity of pyrotechnic compositions.

Cannon – A weapon, larger than a small arm, which ejects its projectile by the action of an explosive. It may be mobile or fixed and includes guns, howitzers and breech-loading mortars.

Cannon Cracker – A large *firecracker.*

Cap – A mechanical or electrical explosive device or a small amount of explosive that may be used for firing an explosive charge.

Synonyms: *Percussion Cap, Detonating Device, Detonator, Blasting Cap.*

Capacitor-discharge Blasting Machine – A blasting machine in which electrical energy, stored on a capacitor, is discharged into a blasting circuit containing electric detonators.

Cap, B.B. – See *B.B. Cap.*

Cap, Blasting – See *Blasting Cap.*

Cap, Blasting, Electric – Electric Blasting Cap: See *Blasting Cap, Electric.*

Cap, Blasting, Ordinary – See *Ordinary Blasting Cap.*

Cap Crimper – A mechanical device used to crimp securely the metallic shell of an ordinary blasting cap, a fuse detonator or igniter cord connector securely to a section of inserted safety fuse. It may be a hand or bench tool.

Also see: *Crimp,* and *Crimping.*

Cap, EBW – EBW Cap: Acronym for Exploding Bridge Wire Cap, and synonym of EBW detonator.

Cap, Electric – Electric Cap: *Electric Blasting Cap* is usually shortened to Electric Cap.

Cap, Electric Blasting – Electric Blasting Cap: See *Blasting Cap, Electric.*

Cap, Fuse – See *Fuse Cap.*

Cap Lamp, Electric – Electric Cap Lamp: See *Lamp, Electric Cap.*

Also see: Safety Lamp.

Cap, Ordinary Blasting – See *Ordinary Blasting Cap.*

Cap, Percussion – See *Percussion Cap.*

Also see: *Cap Type Primer.*

Capped Fuse – A length of safety fuse to which a blasting cap has been attached (crimped). In other words, a capped fuse is the combination of safety fuse and plain detonator.

Capped Primer – See *Primer, Capped.*

Cap, Percussion – See *Percussion Cap.*

Cap Sensitive Explosive – An explosive that is tested to detonate with the use of a specific strength detonator (such as # 8 detonator) is called a cap sensitive explosive, such as dynamite. Blasters use cap sensitive explosives from a safe distance by using fuses and blasting caps, or lead lines and electric detonators. Cap sensitive high explosives are also used in the explosive initiation sequence between a detonator and the main charge.

Compare: *Non-cap Sensitive Explosive.*

Cap Sensitivity – The sensitivity of an explosive to initiation by a *blasting cap*. An explosive material is considered to be cap sensitive if it detonates with a # 8 detonator.

Cap Type Primer – A metal or plastic cap that contains a small amount of initiating explosive, which is readily ignited by shock. It serves as an igniting element in small arms cartridges. The purpose of cap type primer is to kindle small arms cartridges.

Also see: *Percussion Cap.*

Carbamate – Symmetrical diethyldiphenylurea.

Carbide – Chemical compounds consisting of carbon and any other element. Commonly calcium carbide is known as Carbide. Carbides are generally hard materials with metallic conductivity. Some carbides, such as sodium carbide may react explosively on contact with water. Many carbides, such as *calcium carbide* decompose in water to form acetylene. Some carbides, such as Al_4C_3 yield methane on hydrolysis. Besides the hazard of the formation of inflammable gas, another fire hazard of certain carbides is the generation of heat in contact with water. In some cases, the temperature may be raised sufficiently to ignite the gas generated.

Carbide, Cuprous – Cuprous Carbide: Synonym of *Copper Acetylide.*

Carbide of Commerce – Synonym of *Calcium Carbide.*

Carburetor – A contrivance for mixing oil vapor with air in the internal-combustion oil engine vapor preliminary to explosion.

Cardox – A metal tube device using carbon dioxide to produce a blasting effect. In this system, the liquid carbon dioxide, in the cartridge, when initiated by a mixture of potassium perchlorate and charcoal, creates a pressure adequate to break undercut coal.

Carrick Detonator – See *Detonator, Carrick.*

Cartridge – 1. Cartridge means a single unit of ammunition containing the propellant, primer and projectile, designed to be fired from a small *firearm.* The ammunition is contained in, or held together by, a cylindrical shaped case, capsule, or shell of metal, pasteboard, or other material. In other words, it is ammunition consisting of a cylindrical casing containing an explosive charge including a cap or other initiating device and a *bullet* to be fired from a handgun or rifle.

2. A rigid or semi-rigid case containing a blasting agent or an explosive charge for blasting. Such a cartridge is usually wrapped to a predetermined diameter and length. The shape of a cartridge is ordinarily cylindrical one. A soft plastic stick of dynamite, slurry or ANFO is a cartridge.

Also see: *Ball Cartridge*, *Blank Cartridge*, *Center-fire Cartridge*, and *Rim-fire Cartridge.*

Cartridge, Ball – See *Ball Cartridge.*

Cartridge, Blank – See *Blank Cartridge.*

Cartridge, Blasting – *See Blasting Cartridge.*

Cartridge, Center-fire – See *Center-fire Cartridge.*

Compare: *Rim Fire Cartridge.*

Cartridge, Flash – See *Flash Cartridge.*

Cartridge for Weapon – *See Weapon, Cartridge for.*

Cartridge for Weapon, Blank – *See Blank Cartridge for Weapon.*

Cartridge, Ignition – See *Ignition Cartridge.*

Cartridge, Jet Engine Starter – See *Jet Engine Starter Cartridge.*

Cartridge, Oil Well – See *Oil Well Cartridge.*

Cartridge, Power Device – See *Power Device Cartridge.*

Cartridge, Pressure – See *Pressure Cartridge.*

Cartridge, Primed – See *Primed Cartridge.*

Cartridge Punch – A device, made of wood, plastic or non-sparking metal, used to punch an opening in an explosive cartridge to accept a detonator or a section of detonating cord.

Cartridge, Rim-fire – Rim-fire Cartridge: A cartridge in which the primer (fulminate) is located in a rim surrounding its base. In this type of cartridge, primer (fulminate) is integral to the shell casing. When the firing pin strikes, the pin pinches the rim against the chamber and causes the rim to explode and ignite the powder.

Compare: *Center-fire Cartridge.*

Cartridge, Safety – See *Safety Cartridge.*

Cartridge, Signal – Signal Cartridge: Cartridges designed to fire colored flares from 'Very' pistols, etc.

Cartridge, Small Arm – Small Arm Cartridge: Cartridges designed to be fired in arms, including machine guns, of caliber not larger than 19.1 mm. Except in the case of a blank cartridge, it consists of a cartridge case, fitted with a primer, containing a propellant powder charge together with a projectile which may be solid, tracer, lachrymatory or incendiary. It may be arranged in boxes, mounted on belts or placed in clips.

Cartridge, Sporting – Sporting Cartridge: The cartridges include, on the one hand, sporting cartridges (smooth bore) consisting of a cartridge case, with a primer, containing a charge of propellant powder and metal pellets and, on the other hand, cartridges for saloon (shooting gallery) rifles or pistols.

Cartridge Strength – A rating, expressed as a percentage that compares a given volume of explosive with an equivalent volume of straight nitroglycerin dynamite.

Also see: *Bulk Strength.*

Cast, Hexatonal – See *Hexatonal Cast.*

Cast Loading – It is one of the three methods of loading high explosive charges into their containers. Cast loading is accomplished by pouring the chemical explosive as a liquid into the container and letting it solidify. Solid forms of explosives are necessarily press-loaded, or pressed into their container.

Cast Primer – A cast unit of explosive used to initiate detonation in a blasting agent. Ideal example of cast primers is pentolite and composition B.

Cast Booster – See *Booster, Cast.*

Catenary System – A Lightning Protection System (LPS) consisting of overhead wire suspended from poles connected to a grounding system via down conductors. Its purpose is to intercept lightning flashes from the protected area.

Catherine Wheel – A circular-shaped firework.

Synonym: Pinwheel

Cavity Charge – Synonym of *Shaped Charge.*

C.D.B. – Acronym for *Cast Double Base* propellant.

CDF – Acronym for *Confined Detonating Fuse.*

C.E. – Acronym for *Composition Explosive.*

Celluloid – Celluloid is a thermoplastic material made from *cellulose nitrate* and camphor; prepared by dissolving in ether and alcohol or other solvents. When the solvent is dried out, celluloid is left out.

If celluloid is heated to a temperature of 1400 to 1600°C (2500 to 3000°F) without the application of flame, it decomposes quickly, emitting an inflammable vapor. If this vapor is mixed with the correct proportion of air and ignited an explosion is inevitable.

Due to the high percentage of oxygen content in celluloid, it will continue to burn in the absence of air, and for this reason it is nearly impossible to extinguish fires in large quantities of celluloid. Huge quantities of water should be applied for cooling purposes.

In the recent past, all cinematograph film had a cellulose nitrate (celluloid) base. In the present time, most cinematographic film is made of cellulose acetate, and stringent precautions are not required.

Also see: *Safety Celluloid, Collodion.*

Celluloid, Safety – See *Safety Celluloid.*

Cellulose – Cellulose is a compound of carbon, hydrogen and oxygen. The molecule of cellulose is very large and is represented by the formula $(C_6H_6O_5)n$, where n is about 10,000. Cotton, wood fiber or wood pulp and other vegetable fibers consist essentially of cellulose. Such materials are, therefore, used in the production of nitro-cellulose (correct name being cellulose nitrate). However, cotton is almost exclusively used for the production of explosives.

Cellulose Nitrate – Cellulose Nitrate is the correct chemical name of what is popularly known as Nitro-cellulose. It is the nitric ester of *cellulose* and not a nitro compound, having the empirical formula $[C_6H_7O_2\ (ONO_2)_3]x$. It is not a sharply defined compound; rather a range of compounds formed by treatment with a mixture of nitric acid and sulfuric acid of cellulose.

The first great advance in the chemistry of explosives took place in 1846 when Christian Friedrich Schönbein, Professor of Chemistry, University of Basle invented gun cotton, a nitro-cellulose with more than 12.3% nitrogen.

Properties of cellulose nitrate are dependant on the extent to which the cellulose is esterified. Cellulose nitrate is usually formulated as $C_6H_7O_5(NO_2)_3$, $C_{12}H_{14}O_{10}(NO_2)_6$, or $C_{24}H_{30}O_{10}(NO_3)_{10}$.

When ignited unconfined, even the highly nitrated celluloses burn quietly. Since this material is a poor heat conductor and may allow the heat to accumulate, the danger of explosion increases with the quantity burned.

The use for which cellulose nitrate is designed is generally based upon its nitrogen content, which is a direct measure of the degree of esterification.

Although the highly nitrated celluloses are high explosives, these are not suitable for use in their natural light fluffy form. However, they can be compressed by suitable measures to a density as high as 1.4. They are then used as the main charge for submarine depth bombs.

Cellulose nitrate is the universal basis of propellant powders and hence it is of tremendous military importance.

Alfred Nobel first gelatinized cellulose nitrate with nitroglycerin and introduced the first double-base smokeless powder. In gelatinized form cellulose is used as the basis of all propellants, particularly cannon powder or in mixture with guncotton in smaller caliber guns; or mixed with nitroglycerin in double-base powders.

Although it is an explosive, when cellulose acetate is stored wet or in solution in alcohol and contained in airtight drums, it is not regarded as an explosive. It is then regarded as a combustible material, a hazardous substance of class 4.

Synonyms: *Nitrocellulose*; *Dekanitrocellulose*; *Pyrocellulose; Pyrocotton; Pyroxylin; Guncotton; Military Guncotton; Collodion Cotton;* and *Collodion wool.*

Also see: *Collodion, Nitro cellulose, and Smokeless Powder.*

Cellulose Solution – A solution that contains cellulose nitrate and solvent, which gives off flammable vapor.

Center-fire Cartridge – A cartridge in which the primer is located in the center of the base of the shell casing, instead of being contained in its rim. In the case of the Prussian needle gun, the primer is applied to the middle of the base of the bullet.

Compare: *Rim Fire Cartridge.*

Ceyote Blasting – See *Blasting,Ceyote.*

Ceyote Shooting – Synonym of *Ceyote Blasting.*

CH-6 – A finely divided gray powder, which is a mixture of 97.5% RDX, 1.5% Calcium stearate, 0.5% Polyisobutylene and 0.5% Graphite. It is more available and less toxic than tetryl.

Chambering – The blasting term means the process of enlarging a portion, usually the bottom, of blast hole by firing a series of small explosive charges. Enlargement of the blast hole can also be achieved by mechanical or thermal methods.

Charge – A given quantity of explosive.

Also see: the entries hereunder.

Charge, Adobe – See *Adobe Charge.*

Charge, Base – See *Base Charge.*

Charge, Block Demolition – See *Block Demolition Charge.*

Charge, Booster – See *Booster Charge.*

Charge, Bursting – See *Bursting Charge.*

Charge, Cavity – Cavity Charge: Synonym of *Shaped Charge.*

Charge, Commercial Explosive – See *Commercial Explosive Charge.*

Charge, Commercial Shaped – See *Commercial Shaped Charge.*

Charge, Demolition – See *Demolition Charge.*

Charge, Density of – See *Density of Charge.*

Charge, Depth – See *Depth Charge*.

Charge, Detonating – See *Detonating Charge.*

Charge, Expelling – See Expelling Charge.

Charge, Explosive – See *Explosive Charge.*

Charge for fire extinguisher, Explosive Expelling – Explosive Expelling Charge for fire extinguisher: Contrivances containing some propellant explosive with their means of ignition. They are used quickly and completely to expel the extinguishing agent from 'one shot' fire extinguishers.

Charge, Igniter – See *Igniter Charge.*

Charge, Intermediate – Intermediate Charge: Generally intermediate charge is synonym of *booster.* Still in some *explosive trains* an intermediate charge functions between the initial charge and the booster to ensure the detonation of the booster.

Charge, Main Explosive – Main Explosive Charge: It is the explosive material that performs the principal work of blasting.

Charge, Multi-section – See *Multi-section Charge*.

Charge, Normal – Normal Charge: The propelling charge employing a standard amount of propellant designed to fire a small arm under ordinary conditions, as compared with a reduced charge or a supercharge used in special circumstances.

Charge, Priming – See *Priming Charge.*

Charge, Propelling – See *Propelling Charge.*

Charger, Blast Hole – See *Blast Hole Charger.*

Charge, Shaped – See *Shaped Charge.*

Charge, Shaped, Commercial – Commercial Shaped Charge: Cases comprising a charge of detonating explosive with a hollowed-out portion (cavity) lined with rigid material and designed to produce a powerful, penetrating jet effect.

Charge, Shaped, Linear – See *Linear Shaped Charge.*

Charge, Shaped, Flexible Linear – Flexible Linear Shaped Charge: See *Shaped Charge, Flexible Linear.*

Charge, Single-section – See *Single-section Charge.*

Charge, Supplementary Explosive – See *Supplementary Explosive Charge.*

Cheddite – A chlorate mixture class of explosive, the main ingredient of which is potassium chlorate, made up with proportions of dinitrotoluene or nitronaphthalene and from 5% to 6% castor oil.

Chemical Constitution of Explosives – The very fact that some common chemical compounds can undergo explosion when heated indicates that there is something unstable in their structures. While no precise explanation has been developed for this, it is generally recognized that certain groups, nitro dioxide (NO_2), nitrate (NO_3), and azide (N_3), are intrinsically in a condition of internal strain. Increased strain through heating can cause a sudden disruption of the molecule and consequent explosion. In some cases, this condition of molecular instability is so great that decomposition takes place at ordinary temperatures.

Chemical Explosion – The explosion that occurs due to the chemical reaction of a chemical substance or substances may be termed as a chemical explosion. The chemical explosion involves a limited set of simple reactions all of which involve oxidation. It is considered as a very rapid oxidation or combustion. The explosive involved must have its fuel and oxygen within itself. In the chemical compound, the fuel and the oxidation agent are in each molecule, such as TNT; while in the mechanical mixture, the combustible substance (fuel) is mixed with some oxygen supplying substance, such as black powder.

There is one of the two basic phenomena in a chemical explosion, *detonation* or *deflagration.*

The explosion type chemical reactions are: (i) energetically self-sustaining, (ii) high-rate propagation, (iii) exothermic nature, and (iv) characterized by the formation of a large amount of gaseous products.

Chemical explosions may be sub-divided into three principal categories, namely: (i) *anaerobic explosions*, (ii) *aerobic explosions*, and (iii) *decomposition explosions.*

Chemical explosions are the result of human activities.

Chemical Fire Bottle – Chemical fire bottle is in fact an advanced *Molotov cocktail.* Instead of using the burning cloth to ignite the flammable liquid, the chemical fire bottle utilizes the very hot and violent reaction between sulfuric acid and potassium chlorate. Upon breakage of the container, the sulfuric acid sprays onto the paper soaked in potassium chlorate and sugar. The paper, when struck by the acid, instantly bursts into a white flame, igniting the gasoline.

Also see: *Molotov Cocktail.*

Chemical Fuse Bomb – A general-purpose bomb with a chemical fuse. The chemical fuse has been designed to activate upon impact.

Chemical Fuse – A fuse in which substances separated until required for action are brought into contact, and uniting chemically, produce explosion.

Cherry Bomb – A red ball-shaped firecracker with high explosive power.

Chile Saltpeter – Common name for Sodium Nitrate.

Chlorate Mixture – Chlorate mixture is a class of explosives. Ideal examples of chlorates used in the explosive industry are potassium-, sodium- and barium chlorate.

A chlorate is a combination of a metal or hydrogen and the -ClO_3 monovalent radical. The chlorates, being powerful oxidizing agents when mixed with combustible materials, may form explosive mixtures. This class of explosives is usually sensitive and explodes on friction. The other ingredients in the chlorate mixture explosives are combustible materials such as metallic powders, powdered sulfur, powdered charcoal or possibly mixtures of organic matter. Nitro derivatives of benzene, toluene, and other aromatic compounds are also added. Paraffin may be added as a desensitizer.

Chlorates are used in the manufacture of primer caps in combination with mercury fulminates, phosphorus, antimony sulfide and other combustible substance. It is used in pyrotechnic mixtures, as a component of airplane flares and aerial bombs. It is also used as a component of permissible/permitted explosives.

Also see: *Potassium Chlorate.*

Cigar Burning – The burning of a cylindrical charge in a propellant from one end only, the other surfaces being inhibited.

Cladding, Explosive – Explosive Cladding: Synonym of *Explosive Welding.*

Class A Explosives – Classification defined by the U.S. Department of Transportation (DOT). Explosives that possess detonating or otherwise maximum hazard, such as but not limited to, dynamite, nitroglycerine, lead azide, blasting caps and detonating primers.

Class B Explosives – Classification defined by the U.S. Department of Transportation (DOT). Explosives that possess flammable hazard, such as but not limited to, propellant explosives, photographic flash powders, and some special fireworks.

Class C Explosives – Classification defined by the U.S. Department of Transportation (DOT). Explosives that contain Class A or Class B explosives, or both, as components but in restricted quantities.

Classification of Detonator – See *Detonator, Classification of.*

Classification of Dangerous Goods – See *Hazardous Substances, Classification of.*

Classification of Explosives – Explosives differ in a number of characteristics, such as chemical structure, sensitivity, and stability. They can be classified according to different criteria. Some explosive materials can fall into either category, according to how they are initiated. For example, nitrocellulose deflagrates if ignited, but detonates if initiated by a strong detonator. Gunpowder burns if uncontained, but will detonate if contained and fired. However explosives may be classified as follows:

(1) Based on the reaction speed (rate of decomposition), there are basically two categories of explosives, namely:

   (i) High explosives, and
   (ii) Low explosives.

(2) From the standpoint of their chemical composition, explosives may be divided into:
   (i) Single molecule explosive (chemical compound/explosive substance), e.g. RDX, PETN, TNT, etc.; and
   (ii) Composite explosive (mechanical mixture, i.e. mixtures of an oxidizer and a fuel), e.g. ANFO, Gunpowder, etc.

(3) Another classification is based on the sensitivity to initiation, and high explosives may be divided into the following three groups:
   (i) Primary Explosives (e.g. Lead Azide, Mercury Fulminate); and
   (ii) Secondary Explosives (e.g. Dynamite, TNT).
   (iii) Tertiary Explosives (e.g. ANFO)

(4) Classification may be done by the type of explosion; explosives are generally of two types, namely:
   (i) Detonating explosives, and
   (ii) Deflagrating explosives.

(5) Based on their service use a practical classification of explosives is as follows:
   (i) Propellant and impulse Explosives
   (ii) Disrupting or bursting Explosives
   (iii) Initiating Explosives
   (iv) Auxiliary Explosives
   (v) Pyrotechnic Substances

(6) Generally accepted classifications of explosives are as:
   (i) Primary Explosives,
   (ii) Secondary Explosives, and
   (iii) Propellants

(7) According to UN code, Explosive substances are divided into six divisions, based on the degree of hazard. This categorization coincides with US regulations. The divisions are:

   (i) Division 1.1 Explosives with a mass explosion hazard.
   (ii) Division 1.2 Explosives with a projection hazard but not a mass explosion hazard.
   (iii) Division 1.3 Explosives with predominantly a fire hazard and either a minor blast hazard or a minor projection hazard or both, but not a mass explosion hazard.
   (iv) Division 1.4 Explosives with no significant blast hazard.

(v) Division 1.5 Very insensitive explosives that have a mass explosive hazard: blasting agents.
(vi) Division 1.6 Explosive article, extremely insensitive.

(8) According to Mining Engineers, explosives can be classified into two main categories:
(i) Blasting Agents (non-cap sensitive explosives)
(ii) High Explosives (cap sensitive explosives)

(9) From the standpoint of safety, explosives may be categorized as:

(i) Permissible / Permitted explosives; and
(ii) Non-classified explosives.

(10) In the Explosives Act and the Rules now prevailing in Bangladesh, explosives have been classified, in line with the British system, in the manner specified below:

(i) Gun powder;
(ii) Nitrate-mixture;
(iii) Nitro-compound;
(iv) Chlorate-mixture
(v) Fulminate
(vi) Ammunition, and
(vii) Firework.

(11) Based on molecular groupings, explosive substances have been classified by Plets into eight classes, namely:

(i) Acetylides, -C≡C-
(ii) Azides, -N=N- and -N=N=N-
(iii) Chlorates and Perchlorates, $-OClO_2$, and $-OClO_3$
(iv) Fulminates, -N=C
(v) Nitramines, $-NH-NO_2$
(vi) Nitric Esters, $-ONO_2$
(vii) Nitro Compound, $-NO_2$
(viii) Peroxides and Ozonides, -O-O- and -O-O-O-.

(12) From the standpoint of use, explosives may be classified as follows:

(i) Commercial Explosives; and
(ii) Military Explosives.

(13) The most satisfying classification of explosives seems to be one shown in figures that follow:

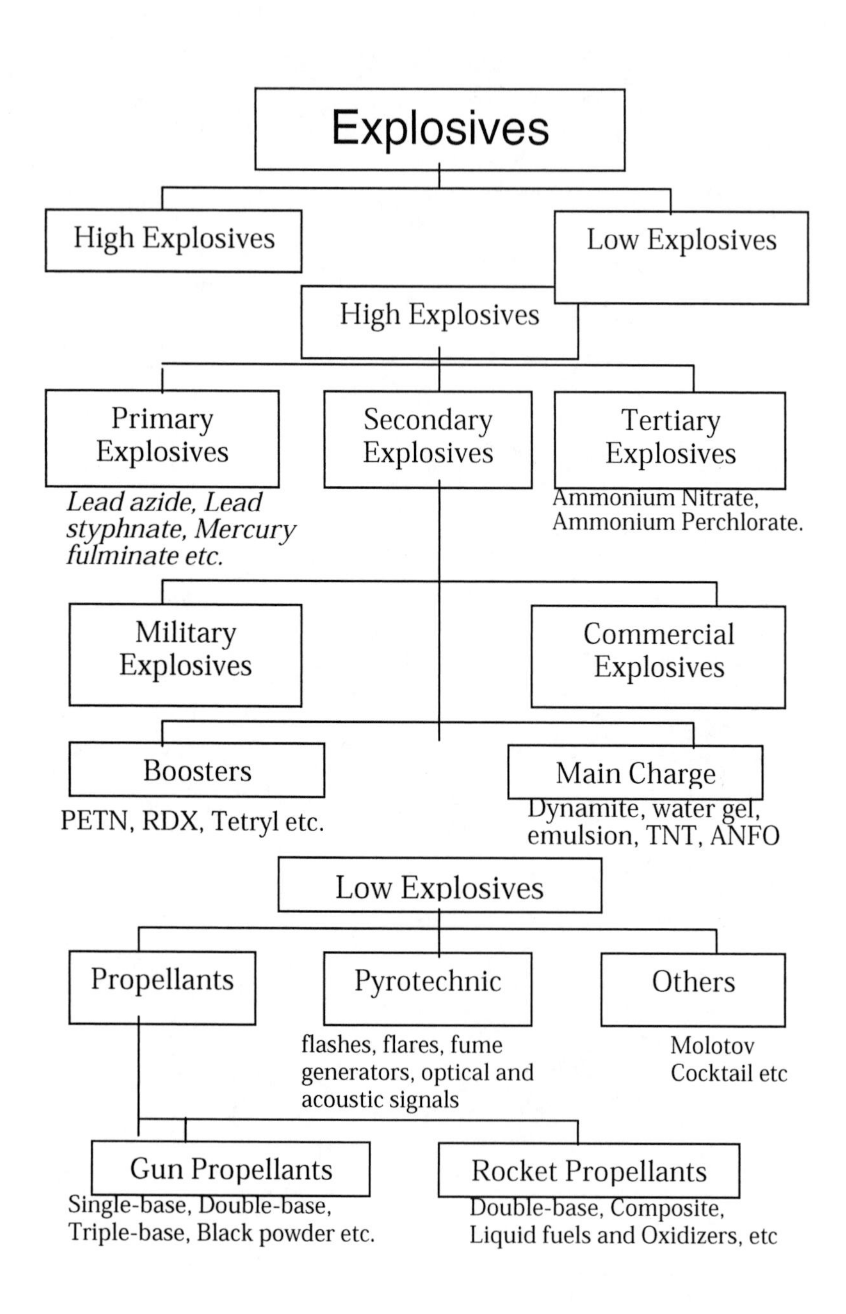
Explosives
High Explosives
Low Explosives
High Explosives
Primary Explosives
Secondary Explosives
Tertiary Explosives
Lead azide, Lead styphnate, Mercury fulminate etc.
Ammonium Nitrate, Ammonium Perchlorate.
Military Explosives
Commercial Explosives
Boosters
Main Charge
PETN, RDX, Tetryl etc.
Dynamite, water gel, emulsion, TNT, ANFO
Low Explosives
Propellants
Pyrotechnic
Others
flashes, flares, fume generators, optical and acoustic signals
Molotov Cocktail etc
Gun Propellants
Rocket Propellants
Single-base, Double-base, Triple-base, Black powder etc.
Double-base, Composite, Liquid fuels and Oxidizers, etc

Classification of Hazardous Substances – See *Hazardous Substances, Classification of.*

Clearing Test – A test for the determination of the speed of solution of nitrocelulose in nitroglycerine.

Closed Bomb – A fixed volume chamber used for testing the pressure-time characteristics of cartridges or combustible materials.

Also see: *Ballistic Bomb.*

Cluster Bomb – A cluster bomb consists of small explosive bomb-lets carried in a large canister that opens in mid-air, scattering them over a wide area.

It is used for soft targets over a wide area. The cluster bomb acts like a shotgun, covering a wider area with a scattering of bomb-lets. A cluster bomb is generally dropped by an aircraft. However, it may be delivered by rocket or artillery projectiles.

The reason behind the development of the cluster bomb is that a single bomb is more useful for soft targets because it covers a smaller area, and its effectiveness is dependent on the accuracy of the bomb's drop. With a view to improving the efficiency of aerial attacks, particularly against soft targets like personnel, cluster bombs were developed.

Also see: *Butterfly Bomb.*

Coal Drill, Electric – A drill driven by electric motor designed for drilling holes in coal for placing blasting charges.

Coal Dust – See *Dust Explosion*, and *Permissible/Permitted* Explosives.

Coal Mine Delay – Coalmine delays are electric detonators designed for use in permissible applications, primarily in underground coalmines. The leg wire insulation and identifying bands are color-coded for identification of each delay period. Usually non-electric systems are not allowed in underground coalmines. Coal mine detonators are normally supplied only with iron or copper-clad iron leg wires to allow for magnetic removal from the coal.

Coal Mine, Gassy – See *Gassy Coal Mine.*

Cocktail, Molotov – See *Molotov Cocktail.*

Cold Zone – The area where equipment and personnel directly support the incident.

Synonym: Safe Zone, and Support Zone.

Collodion – A solution of cellulose nitrate in ethyl alcohol and ether is known as collodion. If this solution also contains camphor, it becomes celluloid when it is dried. It is a tough and hard material. It is used in the manufacture of some toys, imitation tortoiseshell, fancy articles, knife-handles, etc. It is highly inflammable, is easily ignited, and burns rapidly and fiercely.

Collodion Cotton – Collodion cotton is a form of *Nitro-cellulose.* It is white colored pulped fibers. It contains from 11.2 to 12.2% nitrogen. A solution of cellulose nitrate in ethyl alcohol and ether is known as collodion. If this solution further contains camphor it becomes celluloid when it is dried. Collodion cotton is used as gelatinizing agent for *nitroglycerine* in manufacture of *blasting gelatin* and *gelatin dynamite.*

Also see: *Cellulose Acetate*, and *Nitro-cellulose.*

Colored Smoke – See *Smoke, Colored.*

Combustible Gas Indicator – Synonym of *Explosimeter.*

Combustion – Combustion is a chemical reaction, or complex of chemical reactions, in which a substance combines with oxygen to produce heat, light, and flame. It is essentially an oxidation reaction. The combustion reactions that supply most of the energy required by human civilization involve the oxidation of fossil fuels in which carbon is converted into carbon dioxide and hydrogen is converted into water (steam).

According to the nature of combustion, it may be divided into the following classes:

(i) Slow or incipient combustion: one in which the amount of heat and light emitted is feeble.
(ii) Rapid or active combustion: one in which a considerable amount of heat and light are emitted within a short time.

(iii) Deflagration: a case of combustion which takes place with considerable rapidity evolving heat and light.
(iv) Explosion: a very rapid combustion accompanied by a loud report within an extremely short time, with generation of very high pressure and temperature.

Combustion Explosion – Synonym of *Deflagration Explosion.*

Combustion, Spontaneous – See *Spontaneous Combustion.*

Combustion Triangle – See *Fire Triangle.*

Combustion without Heat – Moist yellow phosphorus emits a blush glow of light without evolution of much heat. This is known as Phosphorescence.

Combustion without Heat and Light – Oxidation or rusting of iron, without rise in temperature and without a flame, is an example of combustion without heat and light.

Combustion without Oxygen – The chemical reactions under the definition of combustion, which produce heat and light, although oxygen does not take part in the reactions, such as the burning of powdered metals like iron, magnesium, aluminum, or hydrogen in an atmosphere of chlorine gas.

Commercial Explosive – An explosive that is designed, produced, and used for commercial or industrial applications other than military. Explosives have played an important part in the peaceful progress of civilization. Even in the times of peace, hundreds of thousands of tons of explosives are produced for rockets and fireworks, and for mining, blasting and sporting purposes.

Explosives react rapidly to produce large quantities of gas and heat. Products of such reactions are produced in such a rapid fashion, milliseconds or less, that they provide the opportunity for useful work and become essential tools for the mining, quarrying and construction industries.

Commercial Explosive Charge – Casings comprising a charge of secondary detonating explosive arranged in a multitude of configurations and sizes, used for explosive welding, jointing, forming and other varied metallurgical techniques and processes.

Commercial Shaped Charge – A case which is comprised of a charge of detonating explosive with a hollowed-out portion (cavity) lined with rigid material, and which is designed to produce a powerful, penetrating jet effect.

Common Series – A circuit that is used to connect two or more electric detonators to a single blasting machine.

Compatibility – Compatibility means ability of substances to be stored intimately without occurrence of chemical reaction. Incompatibility may be very hazardous or may result in a loss of effectiveness of the materials. For example, ammonium nitrate explosives and chlorate explosives are incompatible because of the formation of self-decomposing ammonium chlorate.

Comp C – Acronym/Synonym of Composition C.

Complete Round – 1. A complete round of fixed or semi-fixed ammunition consists of a primer, propelling charge, cartridge case and a projectile.
2. A complete round of separate-loading artillery ammunition comprises a primer, propelling charge and (except for blank ammunition) a projectile.

Component, Explosive Train – See *Explosive Train Component.*

Composite Explosive – A composite explosive means the explosive that is formed by two or more constituents, intimately mixed mechanically, in which there must be an oxidizer and a fuel, such as ANFO, Black powder.

Compare: *Composition Explosive.*

Composite Propellant – A propellant based on an oxidizing salt and a fuel/binder, usually not being nitrocellulose and nitroglycerin. Black powder is an example of composite propellant.

Composition A – Composition A explosives consist of a series of formulations of phlegmatized Cyclonite (Hexogen, RDX, Trimethylenetrinitramine) and plasticizing wax. It is a granular explosive. Three widely-known varieties of Composition A are designated as Composition A-1, A-2, A-3. The varieties differ from each other only by the various kinds of wax they contain. Composition A is used as the bursting charge in rockets and land mines, velocity of detonation in confined state being 8,100 meters (26,600 feet) per second at a density of 1.71g /cm^3.

Also see: *Composition Explosive.*

Composition B – Mixtures of approximately 60% Cyclonite (Hexogen, RDX, Trimethylenetrinitramine,) and 40% TNT. Some of them contain wax as an additive. Two varieties of this military explosive are Composition B and B-2. Composition B, when cast, has a density of 1.65g/cm^3, and a velocity of 7600 meters (25,000 feet) per second. It is used as filling for bombs, mines and hollow (shaped) charges. It is also useful as a primer for blasting agents.

Also see: *Composition Explosive.*

Composition B Booster – See *Booster, Composition B.*

Composition C – A military plastic explosive, which is a mixture of Cyclonite (Hexogen, RDX Trimethylenetrinitramine) and a non-explosive plastic agent. This composition explosive is approximately 88.3% RDX and 11.7% plasticizer.

Composition C has greater brisance than TNT; its advantages are that it can be molded into any required shape. The rate of detonation is about 8000 meters (26,000 feet) per second. It is mostly used for bombs in a mixture with TNT (70%), in torpedoes, warheads, and for cutting hard steel. Like tetryl, it can be used as a booster.

Synonym: Comp C.

Also see: *Composition Explosive.*

Composition C-2 – A military plastic explosive, which is comprised of Cyclonite (Hexogen, RDX, Trimethylenetrinitramine) and an explosive plasticizer. The explosive plasticizer was composed of mononitrotoluene. The percentage is approximately 80% RDX and 20% plastic agent.

Acronym: *C-2.*

Composition C-3 – A military plastic explosive consisting of Cyclonite (Hexogen, RDX, Trimethylenetrinitramine) and an explosive plasticizer. This composition explosive is 78% RDX and 22% plasticizer.

Acronym: *C-3.*

Composition C-4 – A common variety of military plastic explosive, which consists of Cyclonite (Hexogen, RDX Trimethylenetrinitramine), a non-explosive plasticizer, a binder and lately marker or taggant chemicals to help detect the explosive and identify its source. RDX makes up approximately 91% of the C-4 by weight. This composition explosive is an off-white solid with the feel of soft clay (putty-like) and has no odor. It is used as block demolition charge as well as other charges. It is effective in temperatures between -70 and +170°F. Velocity of detonation of the explosive is 26,377 ft/sec. C-4 is part of a group of explosives along with Compositin C, C-2, and C-3 each containing different amounts of RDX. It can be shot upon or even thrown into a fire without detonating. The UN now requires that C-4 be odorized upon manufacturing.

Acronym: *C-4.*

Composition, Delay – See *Delay Composition.*

Composition Exploding – Synonym of *Tetryl* in the form of a pellet.

Composition Explosive – The term 'composition' is generally used for any stable explosive. Composition explosives are, in fact, a variety of plastic demolition explosives, which are made from various explosives like RDX and PETN blended with other agents like waxes, oils and plasticizers. The plasticizer itself may or may not be explosive. The blends produce explosives that exhibit special qualities for specific applications. Composition A, Composition B and Composition C are known variants of composition explosives. These are often designated by acronyms like A-2, B-2, C-2 etc. Composition explosives are necessarily castable or moldable ones.

Compare: *Composite Explosive.*

Composition, Illuminant – See *Illuminant Composition.*

Composition, Percussion – See *Percussion Composition.*

Composition, Pyrotechnique – See *Pyrotechnique Composition.*

Concussion Fuse – A fuse ignited by the striking of the projectile.

Confined Detonating Fuse (CDF) – A detonating cord that has a flexible outer sheath to retain the products of detonation.

Confined Detonation Velocity – See *Velocity, Detonation, Confined.*

Confined Explosion – An explosion of a fuel-oxidant mixture inside a closed system such as containers, tanks, rooms. Most rarefied explosions are confined explosions.

If a source of ignition comes in contact in a confined space with a mixture of flammable gas, vapor, dust or aerosol and air in proportions within the flammable limits, the resulting deflagration will liberate heat. The liberated heat will raise the pressure within the confined space ensuring rupture of the confined space.

The violence of the explosion depends upon the nature of the fuel and enclosure containing the mixture as well as on the quantity of mixtures. Explosions occurring from mixtures near the lower or upper limits of the flammable range are less intense than those occurring in intermediate concentrations of the same mixture.

Confined explosions may be subdivided into gas/vapor explosions and particulate explosions, the latter including both dust and aerosol explosions.

A *dense explosion* that occurs in a borehole is also a confined explosion. Some explosives necessarily are physically confined to contain an explosion. This explosion is a confined explosion.

Confinement – Confinement may be defined as an inert material of some strength and having a given wall thickness, situated in the immediate vicinity of an explosive. Priming or heating the explosive materials produces different results, according to whether they are located in a stronger or a weaker confinement. If confined by thick steel, almost any explosive will explode or detonate on being heated; on the other hand, they burn on contact with an open flame if unconfined (Combustion; Mass Explosion Risk), except Initiating Explosives.

The destructive (fragmentation) effect of an explosion becomes stronger if the explosive is confined (stemmed) in an enclosure such as a borehole. In the absence of natural confinement, the explosive charge is often embedded in an inert material such as clay. Mud Cap, Stemming.

Congreve Rocket – See *Rocket, Congreve.*

Contact Operation – See *Operation, Contact.*

Container – Any receptacle, other than a carriage, designed to hold explosives, detonators or primed cartridges.

Control Point – In an explosive operating or test area, the location used for personnel control and operation coordination.

Controlled Explosion – Explosions in the blasting in quarries and mines or in the cylinder of an internal combustion engine are controlled. Very fine industrial-type diamonds used for grinding and polishing are also produced by the carefully controlled use of explosives on carbon.

Contrivance, Water-activated – See *Water-activated Contrivance.*

Conventional Bomb – Conventional Bomb: A category of bombs which are filled with chemical explosives.

Cook-off – When externally applied heat causes the detonation or deflagration of an explosive-filled contrivance the phenomenon is termed cook-off.

Cooling Salt – A flame-depressant, isothermic chemical. Either sodium chloride or sodium carbonate is incorporated in a high explosive to reduce the heat of the explosion as in permitted /permissible explosives.

Cool Explosive – See *Nitroguanidine*, and *Flashless explosive.*

Copper Acetylide – A red-brown solid chemical that contains neither nitrogen nor oxygen and produces no gas on decomposition. It is one of the very few commercial explosives with this characteristic. Evolution of a large amount of heat is the cause of an explosion. Copper acetylide detonates when heated over 100°C or by percussion.

Empirical formula of Copper acetylide is C_2Cu_2 and chemical formula as follows:

```
C —— Cu
||
C —— Cu
```

Synonym: *Cuprous Acetylide* and *Cuprous Carbide*.

Copper Fulminate – A primary high explosive, whose chemical formula is $Cu(ONC)_2$. It is used in the manufacture of detonators to be deployed in gassy coalmines.

Cord, Blasting – S*ee Blasting Cord.*

Cord, Detonating – See *Detonating Cord.*

Cord, Detonating, Flexible – *See Flexible Detonating Cord.*

-cord, Ignita – See *Ignitacord.*

Cord, Igniter – See *Igniter Cord.*

Cord, Splitter – Splitter Cord: See *Ignitacord,* the synonym of splitter cord.

Cordite – Cordite is the designation for double base gun propellants in the United Kingdom. Sir James Dewar (1842-1923) and Sir Frederick Abel invented this important military propellant in 1889. It is composed of nitrocellulose (high-nitrated gun-cotton) and nitroglycerine, containing 30-40% nitroglycerine gelatinized by addition of acetone, thickened and rendered more stable by addition of about 5% mineral jelly (vaseline). The excess of acetone is afterwards evaporated from the gelatinous mass, which is prepared in the form of cords.

It was a rival product of 'Ballistite', an invention of Alfred Nobel.

Also see: *Ballistite*, and *Smokeless Powder*.

Cordtex – British trade name for *Detonating Fuse*.

Cord, Detonating, Metal Clad – See *Metal Clad Detonating Cord.*

Cordeau Detonant – Synonym of *Detonating Cord.*

Cordeau Detonant Fuse – A term used to define detonating cord.

Core Load – The explosive core of detonating cord that is expressed as a number of grains of explosive per 30 centimeters (one foot).

Corning – The stage in the process of black powder manufacture, which produces uniform spherical grains.

Cotton, Collodion – Collodion Cotton: See *Nitrocellulose.*

Cotton, Gun – Gun Cotton: See *Nitrocellulose.*

Cotton, Nitrated – Nitrated Cotton: See *Nitrocellulose.*

Cotton, Nitro – Nitro Cotton: See *Nitrocellulose.*

-cotton, Pyro – See *Pyrocotton.*

Counter – A stage in manufacture of fuse, which winds textile yarns in a direction opposite to a previous spinning process.

Coupling – The term 'coupling' means the degree to which an explosive is in intimate contact with a target surface or fills the cross section of a borehole. In conventional drilling/load shoot operations, bulk-loaded explosives are completely coupled; in the case of linear shaped cutting charges, these are uncoupled; untamped cartridges are decoupled.

Coyote Blasting – A method of blasting in which a number of comparatively large concentrated charge of explosives is placed in one or more small tunnels driven in a rock formation.

Synonym: Coyote Blasting.

Coyote Shooting – Synonym of *Coyote Blasting.*

Cracker – A small firework consisting of a little explosive charge and fuse enclosed in a cylinder-shaped heavy paper casing.

Synonyms: *Firecracker*, and *Snapper.*

Cracker, Cannon – See *Cannon Cracker.*

Crawford Bomb – It is a bomb that is used in the process of determination of burning rates of rocket propellants.

Criminal Bomb – It is an explosive substance that is dropped, placed, projected or thrown with the malicious intention of creating a disturbance, or causing injury or death of persons, or destruction of property.

Crimp – The folded ends of paper explosive cartridges; the circumferential depression at the open end of a fuse cap or igniter cord connector that serves to secure the fuse; or the circumferential depression in the blasting cap shell that secures a sealing plug or sleeve into a electric or non-electric detonator.

Crimper, Cap – See *Cap Crimper.*

Crimping – The act of securing a fuse cap or igniter cord connector to a section of a safety fuse by compressing the metal shell of the cap against the fuse by means of a *cap crimper.*

Critical Diameter – The minimum diameter for propagation of a detonation wave at a stable velocity. It is affected by conditions of confinement, temperature, and pressure on the explosive. It is strongly texture dependent, and is larger in cast than in pressed charges.

Critical Temperature – The temperature above which the self-heating of an explosive causes a runaway reaction. It is dependent on mass, geometry, and thermal boundary conditions.

Crosslinking Agent – The final ingredient added to a *water gel* or *slurry*, causing it to change from a liquid to a gel.

Cuprous Acetylide – See Copper Acetylide, the synonym of Cuprous Acetylide.

Cuprous Carbide – See Copper Acetylide, the synonym of Cuprous Carbide.

Current, All-Fire – See *All-Fire Current.*

Current, No-Fire – See *No-Fire Current.*

Cyanate,Mercury – Mercury Cyanate: Synonym of *Mercury Fulminate.*

Cyclonite – A commercial name for *Cyclo-1,3,5*-trimethylene-*2,4,6*-trinitrimine, better known as RDX, manufactured by the nitration of hexamethylenetetramine [$(CH_2)_6 N_4$]. Empirical formula is: $(CH_2N.NO_2)_3$.

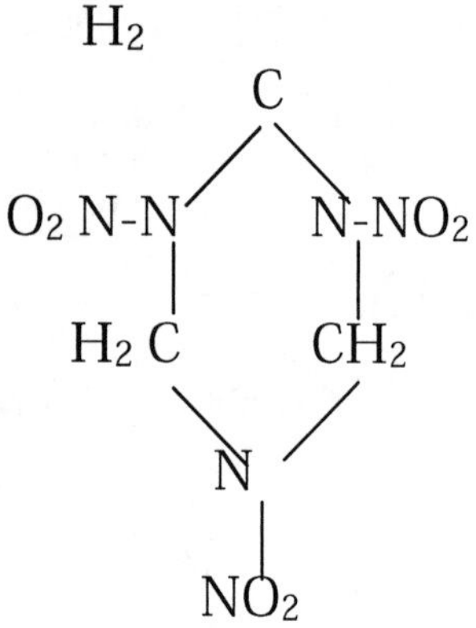

```
        H2
              C
           /     \
O2 N-N              N-NO2
      |              |
   H2 C             CH2
        \          /
            N
            |
           NO2
```

Cyclonite

Cyclonite is a fine white crystalline (rhombic) substance with a high *brisance*. It is odorless, tasteless, non-toxic, non-hygroscopic, and unaffected by light or warm humid air.

Cyclonite is a military explosive. It is one of the most powerful high explosives in use today; it has more shattering power than TNT. It may be used as a *booster* or main *bursting charge*. It is a blasting explosive of high intensity. For structural demolition and in mines and projectile bombs cyclonite is a favored explosive, and it is also used as a propellant when mixed with paraffin wax or with gun-cotton and moderated with various nitro-aromatic compounds.

When combined with proper additives (i.e. plasticizer), it is relatively stable and insensitive to shock. Rate of detonation cyclonite is 7200 meters (26,800 feet) per second.

Synonyms: Cyclotrimethylenetrinitramine, Hexogen, RDX, *T4,* and *Trimethylenetrinitrimine.*

Cyclotol – The name given to RDX/TNT mixtures with compositions varying between 50:50 and 75:25.

Also see: *Composition B.*

Cyclotetramethylenetetranitramine – Chemical name of the substance known by its acronym *HMX.*

Also see: *HMX*.

Cyclotrimethylenetrinitramine – See *Cyclonite.*

Cylinder – A portable closed metal container, the water capacity of which does not exceed one thousand liters, in which compressed gas is held, transported or stored. A cylinder is a small type of *pressure vessel.*

Also see: *Pressure Tank*, and *Pressure Vessel.*

Dangerous Goods – See *Hazardous Substance.*

Danger Zone – The area within which persons would be in physical jeopardy due to overpressure, fragments or firebrands released during shot or round of shots.

Dark Igniter – See Igniter, Dark.

Davy Lamp – Synonym of *Safety Lamp.*

Day Box – A box which is a temporary storage facility, constructed in according to the requirements of safety standards, to be used at the worksite during working hours for storage of explosive materials. The day box is marked with the word 'explosives'.

DDNP – Acronym for *Diazodinitrophenol.*

DDT – An abbreviation for *Deflagration to Detonation Transition.*

Dead Pressed – A highly compressed condition of an explosive, which tends to prevent the *shock to detonation transition* that would otherwise take place. An explosive desensitized by pressurization is termed dead pressed. In dead pressed condition, tiny air bubbles, responsible for sensitivity, are literally squeezed from the explosive mixture.

Dead Pressing – Act of desensitization of an explosive through pressurization. Tiny air bubbles, responsible for sensitivity, are literally squeezed from the mixture by pressurization.

Decomposition Explosion – Decomposition explosion is a type of chemical explosion in which the substance is readily decomposed. Decomposition of lead azide, a primary explosive, is an example in which metallic lead and nitrogen are produced.

Deflagrating Explosive – A deflagrating explosive is an explosive that burns rapidly but does not detonate when used in normal manner. It is an explosive that reacts by deflagration rather than by detonation. Propellants belong to this type of explosives.

Synonyms: *Burning Mixture*, and *Low Explosive.*

Deflagrating Metal Salt of Aromatic Nitro Derivative – These are salts of metal and an acidic aromatic nitro derivative, such as dinitrophenol.

They deflagrate readily on contact with a flame, or as a result of friction. They do not have the characteristics of detonating explosives, such as sodium dinitro orthocresolate, sodium dinitrophenolate, sodium picrate, sodium trinitrocresolate.

Deflagration – Deflagration is one of the two basic mechanisms or types of chemical explosion, the other being *detonation.* Generally, the term 'deflagration' implies the burning of a substance with self-contained oxygen so that the reaction zone advances into the un-reacted material at less than sonic velocity.

The effect of a deflagration under confinement is an explosion. Confinement of the deflagration increases rate of reaction, pressure and temperature and may cause transition into a detonation.

Low explosives (propellants) deflagrate whereas high explosives detonate. However, under certain conditions, high explosives can deflagrate and some propellants can detonate. This happens more often by accident than intentionally. Extremely fine black powder packed too tightly is known to do this.

Unlike detonation, the deflagration rate of an explosive consists of the chemical burning of the material wherein its propagation rates are dependent on chemical kinetics. In this case, heat is transferred from the reacted to the un-reacted material by conduction and convection. Burning rate usually less than 2,000 meters (6500 feet) per second. In deflagration, the output of heat is sufficient to enable the reaction to proceed and be accelerated without input of heat from another source.

Synonym: combustion explosion.

Compare: *Detonation.*

Also see: *Explosive Deflagration* and *Family Tree of Explosions.*

Deflagration and Detonation, difference between – See *Detonation and Deflagration, difference between.*

Deflagration to Detonation Transition (DDT) – In detonators where primary explosives are initiated by heat of flame, but are transformed into a detonation.

Dekanitrocellulose – Synonym of *Cellulose Nitrate.*

Delay – A distinct pause of predetermined time between detonation or initiation impulses, to permit the firing of explosive charges separately. It is in effect a device that causes time to pass from when the device is set up to the time that it explodes.

A delay may be mechanical, pyrotechnic, electronic, or an explosive train component that introduces a controlled time delay in some element of the arming or functioning of a fuse mechanism.

A regular fuse is a delay. Besides fuse delays, there are timer delays and chemical delays. A fuse delay is a very simple means to delay explosive devices that employ fuses for ignition. Timer delays are very often used by terrorists, which may also called time bomb. Although uncommon, chemical delays can be extremely effective in some cases.

Delay-action – Delay-action means detonating some time after the firing. In the case of a projectile it mans detonating some time after the projectile strikes the target.

Delay Arming – It is a *fuse* related term. Delay arming is a device fitted in a bomb or mine to prevent it being actuated for a preset time after releasing or laying. It slows down the arming of the *fuse.*

Delay Blasting – Blasting in which *delay detonators* or *connectors* are used with a view to causing separate charges to detonate at different times, instead of simultaneously. In delay blasting, individual explosive decks, boreholes, or rows of boreholes are initiated at predetermined time intervals using delay detonators.

Antonym: *Instantaneous Blasting.*

Delay Cap, Millisecond – Millisecond Delay Cap: Delay detonators that have built-in time delays of various lengths. The interval between the delays at the lower end of the series is usually 25 milliseconds. The interval between delays at the upper end of the series may be 100 to 300 milliseconds.

Delay, Coal Mine – See *Coal Mine Delay.*

Delay Composition – Delay compositions are mixtures of materials which, when pressed into delay tubes, react without evolution of gaseous products and thus ensure the minimum variation in the delay period. An example of such a mixture is potassium permanganate with antimony.

Delay Connector – A non-electric, short-interval delay device designed for use in delay blasting that is initiated by detonating fuse.

Delay Detonator – A detonator, either electric or non-electric, with a built-in time lag element, which causes a delay between the application of firing signal and the detonation of the base charge.

The delay detonators were first developed in Germany and the UK in the 1930s with a view to help improve rock blasting. The first delay detonators consisted of a small tubular element of aluminum or brass filled with a low burning pyrotechnic between the fuse head and the base charge. The fuse head initiates the pyrotechnic, and when it burns to the other end of the tubular element it detonates the base charge.

Delay time usually ranges from millisecond to a second or more; and the time is based on the length and composition of delay powder.

Delay Detonator, Gasless – See *Gasless Delay Detonator.*

Delay Detonator, Millisecond – See Millisecond Delay Detonator**.**

Delay Electric Blasting Cap – An electric blasting cap in which a predetermined lapse of time, from milliseconds to a second or more, is introduced to delay cap detonation between the application of the firing current and explosion of the base charge. Except for inclusion of delay powder train, it is as instantaneous as an *electric blasting cap.*

Delay Detonator, Zero – See *Zero Delay Detonator.*

Delay Electric Detonator – Synonym of *Delay Electric Blasting Cap.*

Delay Electric Igniter – The expression 'delay electric igniter' means a small metal tube containing a wire bridge in contact with a small quantity of ignition compound.

Delay Element – The device in, or that portion of, a *delay detonator* that produces the predetermined time lapse between the application of firing signal and the detonation of the base charge.

Delay Function – The time or distance interval between the initiation of the fuse and the detonation.

Delay Fuse – A fuse incorporating a means of delaying its action. This type of fuse has a delay element incorporated in the fuse train, permitting the missile to penetrate the target at a distance corresponding to the delay. Such fuses are used to permit penetration of the target before detonation or for mining effect.

In the military, delay fuses are complete shell fuses that set off the explosive charge a definite time after impact.

Delay fuses are classified according to the length of time of the delay.

Delay, Gasless – Gasless Delay: Delay elements consisting of a pyrotechnic mixture that burns without production of gases.

Delayed Inflammation – The term 'delayed inflammation' as applied to *hypergolic* pairs of rocket propellants means the time that elapses from the moment of contact between the reaction partners up to initiation. This delay is of the order of a few milliseconds. The inflammation delay of the reagent pair furfuryl alcohol-nitric acid is about 20 milliseconds.

Synonym: Delayed Initiation.

Delayed Initiation – See *Delayed Inflammation.*

Delay Interval – The nominal time between successive detonations in a blast.

Delay, Long Period – See *Long Period Delay.*

Delay, Millisecond – See *Millisecond Delay.*

Delay Period – A designation assigned to a delay detonator to show its absolute or relative delay time in a given series

Delay Series – A series of delay detonators designed to satisfy specific blasting requirements. There are basically two types of delay series: millisecond (MS) with delay intervals on the order of milliseconds, and long period (LP) with delay times on the order of seconds.

Delay, Small Column Insulated – Small Column Insulated Delay: Slow burning pyrotechnic core contained in a flexible metallic sheath used to produce delay trains.

Delay Tag – A tag, band, or marker on a delay detonator that denotes the delay series, delay period and/or delay time of the detonator.

Delay Time – The lapse of time between the application of a firing signal and the detonation of the base charge of a delay detonator.

Delay, Timer – Timer Delay: Synonym of *Time Bomb*.

Demolition – The term means the breaking up of man-made structures by blasting.

Demolition Charge – An explosive charge contained in a fiberboard, plastic, metal or other material casing, and which is comprised of a secondary detonating explosive.

Dense Explosion – The explosion, which occurs from solids or liquids with explosive properties, is called a dense explosion. The explosions from all commercial explosives are dense explosions. They are characterized by a high level of shattering power or brisance.

Density – The weight per unit volume of explosive, expressed as cartridge count or grams per cubic centimeter or pounds per cubic foot. Density is an important characteristic of an explosive. Raising the density (i.e. by pressing or casting) improves brisance and detonating velocity.

For dynamite, density is expressed as the number of 1-1/4" x 8" cartridges in a 50-pound case.

Density, Bulk – See *Bulk Density*.

Depth Charge – Depth charge is designed to detonate under water and consists of secondary detonating explosives usually contained in metal drums.

Desired Explosion – The intended explosion that occurs or is set to occur in blasting in mines and quarries. Examples of desired explosions are the cylinder of an internal combustion engine, an *explosimeter*, or in the *air bag* of a motor vehicle.

Destructor – Explosive devices designed to destroy missiles or aircrafts or their components.

Detacord – Synonym of *Detonating Cord.*

Deta sheet, (M118), Flex-X – PETN in plastic sheet explosive form. Rate of detonation: 6,800 meters (22,300 feet) per second.

Detcord - Synonym of *Detonating Cord.*

Detection of Explosive – Detection of explosives is essential in order to combat terrorists using explosives, and to locate landmines and unexploded ordnance (UXO). There are a number of methods for detection of explosives, including canines, chemical sensor, lasers, and neutron beam.

Deterrent – A material used for reducing the initial rate of burning. It is applied as a coating on grains of powder.

Detonability – The quality or state of being detonable.

Detonable – Capable of being detonated.

Detonant, Cordeau – Cordeau Detonant: Synonym of *Detonating Cord.*

Detonant Fuse, Cordeau – Cordeau Detonant Fuse: A term used to mean *detonating cord.*

Detonatable – See *Detonable.*

Detonate – To undergo *detonation.*

Also see: Detonation.

Detonating Agent – A sensitive explosive such as fulminate of mercury used to set off another less sensitive high explosive. A detonating agent is a primary high explosive.

Synonym: Detonating Charge.

Detonating Charge – Synonym of *Detonating Agent.*

Detonating Cord – It is a strong flexible cord containing a center core of high explosives, usually *PETN.* The high-velocity explosive core, ranging from 1.5gm to 100gm/meter is protected against moisture and wear by layers of textile and covered with a waterproof plastic coating.

It can serve both as an initiator and as an explosive charge. In some ways it may be considered to be a high-speed fuse that explodes instead of burning and is suitable for detonating high explosives. It may also be termed as a 'very long detonator'. It is used primarily to initiate a series of charges of explosives in boreholes or under water. It can be used under water if the ends are sealed against the ingress of water. In order to initiate several charges, branch cords are attached to a 'main cord'.

A detonator is required to initiate a length of detonating cord. When detonated, it explodes practically instantaneously throughout its length, detonation speed being in excess of 7000 meters (23000 feet) per second. A detonator is connected to one end of a length of detonating cord which, when initiated, causes the instantaneous detonation of the detonating cord and of the other cap-sensitive explosive charges attached to it.

Detonating cord, also called detonating fuse, can serve two functions: (i) simultaneous detonation of several charges avoiding multiple plain or electric detonators; and (ii) continuous initiation of the full length of an explosive column in a blast hole instead of point initiation with individual detonators. It has some advantages over detonators because it is versatile, safe for use in extraneous electricity environments, can be simultaneously fired without detonators, has no hole limit, is totally consumed, provides for sequential blasting through incorporation of delay connector and is low in cost.

Low-yield detonating cord can be employed directly as a charge in precision cutting charge to remove cables, pipes, wiring, fiber optics, and other utility bundles by placing one or more complete wraps around the target. High-yield detonating cord can also be used to cut down small trees; one complete wrap per foot (30 centimeter) diameter is a rough starting point.

Detonating cord (fuse) serves to initiate blasting charges; the initiation is safe if the cord is coiled several times around the cartridge.

Owing to its high-speed detonation, it is possible to synchronize multiple charges simultaneously, even if the charges are placed at significantly varying distances from the primer charge.

A sequential firing can be done by insertion of detonating *relays (delays)* in the firing line. This is particularly required in the case of demolitions, when structural elements need to be destroyed in specific order to control the collapse of a building.

Synonyms: Cordeau, Cordeau Detonant, Cordtex, Detacord, Detcord, Det Cord, Detonating Fuse, Detonation Cord, Explosive Cord, Prima Cord, Primacord, and Primer Cord.

Also see: *Detonating Fuse,* and *Primadet.*

Detonating Cord Downline – The section of detonating cord that extends within the borehole from the ground surface down to the explosive charge.

Detonating Cord, Flexible - Flexible Detonating Cord: It consists of a core of detonating explosive enclosed in spun fabric with or without plastics or other covering and wire countering.

Detonating Cord, Metal Clad – Metal Clad Detonating Cord: It consists of a core of detonating explosives clad by a soft metal tube with or without a protective covering.

It is called 'mild effect detonating cord (fuse)' when the core contains a sufficiently small proportion of explosive.

Detonating Cord, Miniaturized – See *Miniaturized Detonating Cord.*

Detonating Cord MS Connector – A non-electric, short-interval (millisecond) delay device designed for use in delaying blasts that are initiated by detonating cord.

Detonating Cord, Shielded Mild – Shielded Mild Detonating Cord: *MDF* contained in small diameter steel tubing.

Also see: *MDF.*

Detonating Cord Trunk Line – The line of detonating cord that is used to connect and initiate other lines of detonating cord.

Detonating Delay – Synonym of Detonating Relay. See *Relay, Detonating.*

Detonating Device – An electrical or mechanical explosive device or a small amount of explosive, which is designed to use for firing an explosive charge

Synonyms: Cap, and Detonator.

Also see: Blasting Cap, Explosive Device, and Percussion Cap.

Detonating Explosive – An explosive that reacts by detonation rather than by deflagration when used in its normal manner. Detonating explosives are high explosives.

Detonating Fuse – A fuse containing a high explosive. It is a device to detonate the high explosive bursting charges of projectiles, mines, bombs, torpedoes, and grenades.

Synonyms: Cordeau, Cordeau Detonant, Cordtex, Detacord, Detcord, Det Cord, Detonating Cord, Detonation Cord, Explosive Cord, Prima Cord, Primacord, and Primer Cord.

Also see: *Detonating Cord, Fuse,* and *Primadet.*

Detonating Fuse, Confined – Confined Detonating Fuse: A detonating cord having a flexible outer sheath to retain the products of detonation.

Detonating Fuse, Mild – Mild Detonating Fuse (MDF): More accurately called miniature detonating fuse. It is a flexible metal tube, usually lead, containing a much smaller core of high explosive than the normal detonating fuse (cord).

Detonating Fuse – See Fuse, Detonating.

Also see: *Fuse.*

Detonating Fuse, Point – See *Point Detonating Fuse.*

Detonating Gas – A mixture of hydrogen and oxygen in a volume ratio of 2:1, the volume ratio required to form water. It is extremely explosive when ignited.

Detonating Primer – As used in transportation purposes, the term 'detonating primer' is assigned to the device consisting of a detonator and an additional charge of explosives, assembled as a unit.

Detonating Relay – See *Relay, Detonating.*

Detonating Wave – The *shock wave* set up when a detonator is ignited.

Detonation – 1.The term 'detonation' is used to describe an explosion phenomenon of almost instantaneous decomposition. It is an exothermic chemical reaction that propagates through the reaction zone toward the un-reacted material at a supersonic velocity, forming a propagating shock wave. Thus, a detonation may be defined as an explosion process of supersonic velocity involving a sustained shock wave.

Normally a detonation is brought about by a shock wave traveling at supersonic velocity through the material. This shock wave initially compresses the material, heating it and causing it to break down into its constituent molecular materials. This causes the liberation of the fuel and oxidizer that heretofore were chemically bonded, and they recombine to form gases. The formation of gaseous chemical products releases enormous amounts of energy in the form of heat, light, etc. in just billionths of a second, which sustains the shock wave. The almost instantaneous reaction produces a blast of rapidly expanding hot gases.

Some unique events take place in the brief instant of a detonation, such as the detonation wave traveling as fast as 9 kilometers (30000 feet) per second. Power approaches 20 billion watt per square centimeter, the shock wave produces pressure up to 500,000 times that of Earth's atmosphere, and temperatures can soar to 5,500 Kelvin.

In the detonation of an explosive, the rate of reaction is controlled by thermodynamic considerations. In the deflagration of an explosive, the rate of reaction is dependant on chemical kinetics.

Generally, primary and secondary explosives detonate and low explosives deflagrate. However, under certain conditions, low explosives can detonate and primary or secondary explosives can deflagrate. Probably this happens more often by accident than design, and may have disastrous results.

Primary explosives can be detonated from a simple source such as a flame, spark or impact. Secondary high explosives must have sufficient stimulus to initiate detonation. This is usually in the form of an additional explosive charge that sets up a shock wave. A few of these materials also require a booster charge in addition to an initiator to cause them to detonate. Detonation is normally characterized by a crater, if the material is located on or near the surface of the ground.

2. Also used loosely, but incorrectly, to describe the combustion reactions that occur during *knocking* or 'pinking' in an *internal-combustion engine.*

Also see: *Initiation Sequence,* and *Firing Train.*

Compare: *Deflagration.*

Detonation and Deflagration, difference between – Difference between detonation and deflagration: High explosives undergo detonation and low explosives deflagrate.

A detonator initiates detonation, whereas an igniter or a squib initiates deflagration.

The detonation of an explosive is the propagation of a shock wave through the explosive material, where the rates of reaction are controlled by thermodynamic considerations. The deflagration rate of an explosive consists of the chemical burning of the material wherein its propagation rates are dependent on chemical kinetics.

Detonation velocities lie in the approximate range of 1500 to 9000 meters (5000 to 30000 feet) per second, but the rate of deflagration is below that of the detonation velocity. The rate of reaction in a detonation is supersonic, whereas the rate of reaction in a deflagration is subsonic.

In order for a low explosive to explode, it must be contained in a strong enclosure, whereas a high explosive can be detonated without being enclosed.

Detonation Cord – Synonym of *Detonating Cord.*

Detonation, High Order – See *High Order Detonation.*

Detonation, Low Order – Low Order Detonation: A chemical reaction in a detonable material in which the reaction front advances with a velocity that is appreciably lower than the characteristic detonation velocity for the material.

Detonation, Mass – See *Mass Detonation.*

Detonation Pressure – Detonation pressure is the pressure that is produced in the reaction zone of a detonating explosive. It is a function of the explosive's density and the square of its velocity.

Detonation Rate – Velocity at which a detonation wave proceeds through an explosive.

Detonation, Selective – See *Selective Detonation.*

Detonation, Sympathetic – Sympathetic Detonation: The detonation or explosion of an explosive material by the detonation of another charge in the vicinity without actual contact. A sympathetic detonation may be defined as the initiation of an explosive charge without a priming device as a result of receiving an impulse from another explosion in the neighborhood through air, earth, or water. It can be said that the second explosion, known as sympathetic, is initiated by influence.

The distance through which transmission of detonation may occur depends on many factors, such as mass, physical and chemical characteristics, detonation properties of donor charge, existence and characteristics of acceptor charge confinement, existence and characteristics of the medium between the charge, etc. Other relevant factors being constant, the possibility of the propagation of detonation is determined mostly by both the sensitivity of the acceptor charge and by the initiating strength of the donor charge. Possible factors that contribute to sympathetic detonation also include the initiation of acceptor charge by the passage of shock wave from one mass to the other, by flying fragments and the like.

Synonyms: Sympathetic Explosion; Propagation of Detonation, Propagation of Explosion, and Transmission of Detonation.

Detonation System, Gas – See Gas Detonation System.

Detonation, Transmission of – Transmission of Detonation: Synonym of *Sympathetic Detonation.*

Detonation Velocity – The velocity at which a detonation progresses through an explosive. It is the rate of propagation of a detonation in an explosive. It is a characteristic of each individual explosive.

Detonation Velocity, Confined – S*ee* Velocity, Confined Detonation.

Detonation Wave – See *Wave, Detonation.*

Detonation Zone – *See Zone, Detonation.*

Detonative – Exploding almost instantaneously.

Detonative Explosive – Explosive which is characterized by detonation or which possess the property of detonating.

Detonator – A device designed and used to explode, and initiate less sensitive secondary or tertiary high explosives.

A detonator plays the same role in blasting that a spark plug plays in starting an engine.

The term 'detonator' describes a device that contains an explosive, not the explosive itself. It is a complete explosive initiating device, which includes the active part of the assembly, usually enclosed in a metal shell, and the attached initiation signal transmitter such as wires, shock tube, or other signal-transmitting material.

Usually, a detonator consists of a small thin-walled tube made of aluminum or copper or other material, filled with a small quantity of an extremely sensitive primary *explosive* (called *priming charge*) and a small amount of secondary explosive (called *base charge*). Copper tube detonators are mandatory in gassy and/or ignitable dusty underground mines.

Either the flame of a safety fuse or shock tube or an electrical charge creates the initiation (i.e., firing of priming charge) of a detonator. There are two basic types of detonators, both of which may be constructed to detonate instantaneously, or may contain a delay element:

(a) Plain detonators, which are activated by such means as safety fuse, other igniferous device or flexible detonating cord.

(b) Electric detonators, which are activated by an electric current.

Detonators can be initiated by the flame of a safety fuse, non-electrically as in the case of *Shock Tube, or* electrically.

The credit of the invention of the detonator goes to Alfred Nobel who in 1863 invented *practical detonator.* With the development of *practical detonator,* hazardous ignition by safety fuse and black powder was overcome. Nobel made experiments with design, and upon several modifications, eventually created the detonator in 1865 that was comprised of a metal 'blasting cap' charged with mercury fulminate rather than gunpowder. Mercury blasting cap opened the door for all subsequent high explosive use. All detonator advancements are based on original mercury blasting cap.

The strength of detonators is dependent on the amount of base charge they contain and identified by strength numbers ranging from Nos. 1 to 8. The strength increases with the quantity of base charge, and so the number. A No.8 detonator is the highest strength, and contains 0.8 g of base charge, and 0.3 g priming charge. For all practical purposes, No. 8 is the main type of detonator.

Synonyms: Blasting cap, Cap, and Detonating Device.

Also see: *Classification of Detonators*, *Explosive Device, Percussion Cap, Electric Blasting Cap, Delay Electric Blasting Cap, Non-electric Blasting Cap, Non-electric Delay Blasting Cap, Seismic Detonator,* and *Detonating Cord.*

Detonator Assemblies for Blasting – Non-electric detonators assembled with and activated by such means as detonating cord, safety fuse or shock tube. The basic detonators may incorporate delay elements or detonating delays or may be of instantaneous type.

Synonym: Blasting Cap Assemblies.

Detonator, Briska – Briska detonator: The detonator whose base charge is extreme heavily pressed to achieve extra power.

Detonator, Carrick – Carrick detonator is a kind of permitted detonator for use in gassy coalmines. The Carrick detonator was developed to avoid the problem of the ignition of *firedamp* at the time of burning pyrotechnic composition of normal delay detonators. It is made by extruding several lead tubes in a sandwich to give several smaller burning tubes, each tube being too small and cool to kindle firedamp.

Detonator, Classification of – Detonators are generally of three kinds:

1. Plain detonators, which are activated by such means as safety fuse, other igniferous device or flexible detonating cord.
2. Electric detonators, which are activated by an electric current.
3. Electronic detonators, which are activated by an Integrated Circuit or ASIC ("Application Specific Integrated Chip")

Detonators may be called either *instantaneous detonator,* or *delay detonator.*

Detonators may also be categorized with respect to initiation signal energy source, i.e. non-electric, electric, electronic.

Detonators may be classified as percussion, stab, electric, or flash according to the method of initiation.

Depending on design, detonators are activated by heat, shock wave or electric spark, and can be initiated by the flame of a safety fuse, non-electrically (as in the case of *Shock Tube), or* electrically.

Detonator, Delay – See *Delay Detonator.*

Detonator, Delay, Short – See *Short Delay Detonator.*

Detonator, Delay Electric – See *Delay Electric Detonator.*

Detonator, Delay, Millisecond – See *Millisecond Delay Detonator.*

Detonator, EBW – EBW Detonator is an electric detonator wherein the primary explosive is substituted by an EBW. The shock energy from the explosion of the wire directly initiates the base charge. An exploding bridge wire detonator functions instantaneously. This detonator can only be initiated upon application of a high current through the device bridge wire. It cannot be initiated by any normal shock or electrical energy. It is initiated by capacitor discharge. Luis Alvarez developed this detonator in 1940s for the Manhattan project.

Detonator, Electric – A detonator designed and capable of initiation by means of an electric current. It requires electrical energy to activate the explosive train, detonating the base charge.

A detonator for firing by electric current (electric detonator) consists of two insulated leg wires, ignition mixture of lead azide or mercury fulminate, high-resistance bridge wire and a match head. It is detonated via an *exploder*, making initiation easy and safe.

It is similar to a non-electrical detonator except that it is initiated by the application of electrical current through electrical wires. The current causes a bridge wire or match elements to heat/function, causing the ignition charge to explode, which in turn causes a chain reaction to initiate the base charge. The wires are secured into the detonator by a closure plug, crimped into the shell, which seals the explosive from moisture.

Electric detonators may be of a number of types: (a) Electronic delay, (b) Short period delay (c) Long period delay, (d) Seismic, and (e) Instantaneous.

Electric detonators have a number of advantages over plain detonators, such as circuit testing, control of initiation time, higher degree of safety, removal of blaster from shot, suitable for use in underground gassy cold mines. However, there are also disadvantages. An electric detonator has the risk of premature detonation due to extraneous sources of electricity such as lightning, static stray currents and radio frequency energy due to the presence of electrical wire.

Detonator, Electric Delay – Electric Delay Detonator: Synonym of *Delay Electric Blasting Cap.*

Also see: *Delay Detonator.*

Detonator, Electric Instantaneous – See *Instantaneous Electric Detonator.*

Detonator, Electronic – This initiating device is of recent development. The idea for the electronic detonator evolved in the 1990s. Electronic detonators are still in the testing stage. Electronic detonators can fulfill the demand for increased accuracy, but the costly technology has hindered its expected growth.

In an electronic detonator, delay is achieved electronically. A computer chip is used to control delay timing. An integrated circuit chip and a capacitor internal to each detonator control the initiation time.

An electronic detonator has a number of advantages, such as higher precision, improved blasting result due to wide range of delay, reduction of air blasts/ground vibration, and safe use in extraneous electricity environment. Electronic detonators also have some disadvantages, such as enhanced cost per detonator and training for the user.

Detonator, Electronic Delay – Electronic Delay Detonator: See *Electronic Detonator.*

Detonator, Exploding Bridge Wire – See *Detonator, EBW.*

Detonator, Fire – Fire Detonator: It is a synonym of *plain detonator.*

Detonator for Ammunition – A detonator that is specially designed for the initiation of ammunition.

Detonator for Blasting – Small metal or plastic tubes containing explosives such as lead azide, PETN, or similar explosive. There are essentially two kinds, both of which may be constructed to detonate instantaneously, or may contain a delay element:

(a) Non-electric detonators, which are activated by such means as safety fuse, other igniferous device or flexible detonating cord.
(b) Electric detonators, which are activated by an electric current.

Detonator, Fuse – See *Fuse Detonator.*

Detonator, Gasless Delay – See *Gasless Delay Detonator.*

Detonator, High Tension – See *High Tension Detonator.*

Detonator Insensitive Explosive – Synonym of *Blasting Agent.*

Detonator, Instantaneous – Instantaneous Detonator: See *Zero Delay Detonator,* the synonym of Instantaneous Detonator.

Detonator, Instantaneous Electric – See *Instantaneous Electric Detonator.*

Detonator, Millisecond Delay – See *Millisecond Delay Detonator.*

Detonator, No. 8 – The strength of detonators is dependent on amount of base charge contained and identified by strength number ranging from Nos. 1 to 8. The strength increases with the increase of the quantity of base charge, and so the number. A No. 8 detonator being the highest strength one contains 0.8 g of base charge, and 0.3 g priming charge. For all practical purposes, No. 8 is the main type of detonator.

Also see: *Detonator.*

Detonator, Non-Electric – Non-electric Detonator: A detonator in which the wire of the electric detonator is replaced by a thin plastic tube. This typically contains a small amount of the powdered military explosive cyclotetramethylene tetranitramine (HMX) and fine aluminum powder, which can sustain a burning flame from one end and spit it onto the delay element inside the detonator.

Nitro Nobel in Sweden developed non-electric detonator in the 1970s when the company simplified the process of connecting explosives for complicated blasting operations.

Non-electric detonators have the advantage that they can readily be connected by a series of plastic clips. Further, they are not subject to problems of stray electric current and current leakage, which can affect electric detonators.

Detonator, Ordinary, Reinforced – See *Reinforced Ordinary Detonator.*

Detonator, Plain – Plain Detonator: Plain detonator is a flame fired initiating device. It is designed to fire by means of a safety fuse inserted in the shell's open-end and it is used to initiate dynamites, detonating cord, emulsions and other cap sensitive explosives, in a wide variety of applications in mining and civil works. It consists of a cylindrical aluminum shell with one open-end loaded with a primary and a secondary explosive charge, the open-end is crimped to hold the safety fuse in place and to prevent water from getting into the detonator. It is also a non-electric detonator.

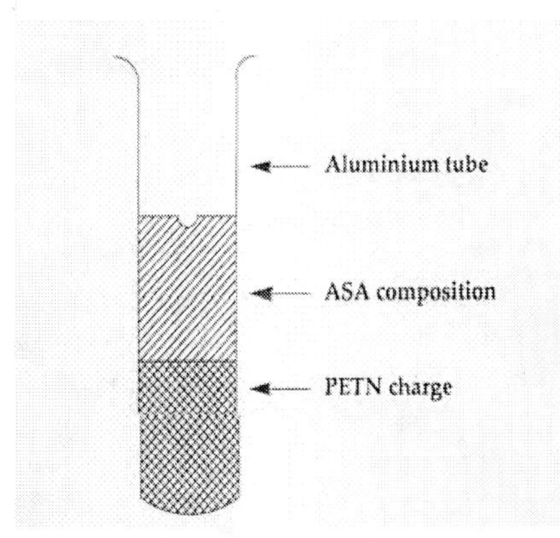

Plain Detonator (Courtesy: ICI)

Detonator, Practical – See *Practical Detonator.*

Detonator, Primer – See *Primer- Detonator.*

Detonator, Reinforced Ordinary – See *Reinforced Ordinary Detonator.*

Detonator, Seismic – Seismic detonator: It is a specific type of instantaneous electric detonator that is made for use in geophysical surveying applications. The seismic detonator is designed to function in less than one millisecond following application of the recommended level of current, and is designed for rugged field usage. They typically incorporate a time break function in which the circuit is interrupted only at the moment of detonation, providing a highly accurate zero time on the seismic recording trace.

Detonator, Semi-conductive Bridge – Semi-conductive Bridge Detonator: In this technology, the bridge wire is replaced by a microchip containing a semi-conductive bridge that flashes upon the application of a low current. This design, when combined with microelectronic circuits, can be used to produce electronic delay detonators.

Detonator Sensitive Explosive – A detonator-sensitive explosive is a secondary high explosive.

Antonym: *Blasting Agent.*

Detonator, Shock Tube – See *Shock Tube Detonator.*

Detonator, Short Delay – See *Short Delay Detonator.*

Detonator, Slapper – See *Slapper Detonator.*

Detonator, Thermal – See *Thermal Detonator.*

Detonator, Zero Delay – See *Zero Delay Detonator.*

D, Explosive – See *Explosive D.*

Diameter, Critical – See *Critical Diameter.*

Diameter, Permissible – See Permissible *Diameter.*

Diazodinitrophenol – A high explosive. It is a yellowish brown powder. Diazodinitrophenol (DDNP) is a less sensitive high explosive to impact than either lead azide or mercury fulminate. It is much less sensitive to friction than mercury fulminate and a more powerful initiator, similar to lead azide in this respect. Heating can cause it to explode. It is desensitized by immersion in water. A cold solution of sodium hydroxide can be used to destroy it.

DDNP is used in commercial detonators as a substitute for mercury fulminate. It is often used as an initiating explosive in propellant primer devices.

Acronym: DDNP.

Synonym: Dinol.

Dinitrobenzol – The single molecule high explosive is used to fill artillery shells and in bursting charges. It is also a useful industrial explosive when mixed with more powerful explosives or with oxygen carriers, such as ammonium nitrate, inorganic chlorates or nitrates.

Dinitroglycol – See *Glycoldinitrate*, the synonym of Dinitroglycol.

Dinitromonochlorhydrin – This single molecule high explosive is used as a component in low-freezing dynamites. It is also used as a component in permissible gelatinous explosives. Dinitromono-chlorhydrin is miscible with nitroglycerine.

Dinitronaphthalene – This compound is a weak high explosive but it has toxic hazard. Dnitronaphthalene is used mixed with chlorates and perchlorates and in combination with picric acid. It can be used as an ingredient of some permissible explosives and in combination with ammonium nitrate.

Dinitrophenol – The single molecule high explosive, which is used as a component of some shell and bomb charges. One such mixture is with picric acid. 2,4-Dinitrophenol is used to lower the melting point of straight *picric acid.*

Dinitroresorcinol – This acid compound and its lead salt are used in commercial priming compositions and blasting caps. It is used to facilitate the ignition of lead azide.

Synonym: Styphnic acid

Dinitrotoluene – 2,4-Dinitrotoluene is a toxic substance and hazardous to use. It is used as an ingredient of flashless nonhygroscopic and nonhygroscopic smokeless powders. Dinitrotoluene is used as a deterrent and cooling agent and as a gelatinizer for propellant powders. It is also used as an ingredient in some permissible and chlorate explosives.

Dinol – A commercial term for *Dinitrodiazophenol (DDNP).*

Di-Oil – Commercial name for Dinitrotoluene.

Direct Contact with Explosive – Physical touch of an electrical instrument or equipment with bare explosives, the metallic casing of an explosive, or the firing leads of an explosive device.

Dirty Bomb – A conventional bomb that contains biological, chemical, or radiological agents. Generally dirty bomb means a conventional explosive impregnated or wrapped with radioactive substance, the release of which is intended to contaminate an area or population. The term dirty bomb is most often used to refer to a Radiological Dispersal Device (RDD), a radiological weapon which combines radioactive material with conventional explosives. Combining the radioactive material as used in industry (contrary to one for medical use) with a dispersion device is quite capable of rendering an area uninhibited. It is a much easier device in comparison to a nuclear bomb.

Synonym: Explosive Dispersion Device.

Dispersive Bomb – A categoriy of bombs which are filled with submunitions, chemicals or other disruptive agents that spread on impact.

Disposal of Explosives – It means the destruction or disposal of deteriorated or unserviceable explosives or of the explosives waste. There are a number of means of disposal of or destruction of an explosive. The method used is dependant on the nature of the explosive and its hazards, and the type and position of the disposal site. There are four recognized ways of disposal or destruction of explosives, namely:

(1) burning;
(2) chemical destruction.
(3) detonation; or
(4) dissolving or diluting by a solvent.

Utmost care and precaution should be taken for disposal or destruction of explosives. One of the main causes of accidents with explosives is the disposal of explosives. Explosives earmarked for destruction should be treated as unusually unstable due to deterioration. A properly considered system of work with appropriate safety precautions has to be drawn for safe disposal.

The authority having jurisdiction should be informed beforehand of the planned disposal and their advice should be taken.

Disrupting Explosive – Explosives employed with the aim of creating damage to the target under attack. These are high-explosive charges that are used alone or as a part of the explosive charge in projectiles as a bursting charge, and in mines, bombs, depth charges, missiles and torpedo warheads.

Synonym: Bursting Explosive.

Ditching Dynamite – See *Dynamite, Ditching.*

Dobying – Synonym of *mud-capping.*

Donor – An exploding charge producing an impulse that impinges upon an explosive 'acceptor' charge.

Dope – Individual, dry, non-explosive ingredient that comprises a portion of an explosive formulation.

Double Base Propellant – See *Propellant, Double Base.*

Down Conductor – A form of a main conductor designed to conduct the current of a lightning flash vertically down to the earth electrode system.

Drift – An underground tunnel through stone in coal mining.

Drill Ammunition – See *Ammunition, Drill.*

Driver – Driver is a small unit similar to an *explosive switch* in which a piston is pushed forward by a small explosive and/or propellant charge.

Dry Blasting Agent – See *Blasting Agent, Dry.*

Drogue Gun – See *Gun, Drogue.*

Dual Use Explosive – An explosive or explosive product that can be utilized for both civil and military purposes.

Dud – An explosive device that has failed to initiate as intended.

Dummy Projectile – See *Projectile, Dummy.*

Duration – The term 'duration' is a shortcut to the term 'positive phase duration', which means the time taken for the pressure pulse to decline to zero.

Dust Explosion – A dust explosion occurs when a finely divided solid carbonaceous, or other combustible material is held in suspension in the air, forming a flammable cloud through which a flame propagates. However, even materials generally not considered combustible can produce an explosion when burned in suspension.

The combustion reaction is a surface reaction; the rates of pressure rise generated by combustion are largely dependent on the surface area of the dispersed dust particles. The finer the dust, the more violent the reaction.

A dust explosion has both similarities and differences with a gas or vapor explosion. Indeed both types of explosions fall under the general category of *rarefied explosions.* The most important difference is the heterogeneous character of dust explosions. This contrasts with the homogenous nature of gas or vapor explosions, in which the reactants are mixed on a molecular level.

When the flammable dust suspended in the air in the form of a cloud it becomes hazardous, because it is then in intimate contact with the oxygen essential for combustion. The dust particles can absorb oxygen like a sponge. The finer the state of division the more dangerous the dust, owing to the much greater surface exposed to the air, and the longer the period during which it will remain in suspension. In this state contact with a suitable source of heat will result in an explosion. The maximum effect is obtained when the proportions of dust to air are such that there is just sufficient oxygen to ensure complete combustion. For a dust explosion, the source of ignition usually must be larger than that necessary for gas-air mixtures.

Very often dust explosions are of greater intensity than gas explosions, because of the fact that the larger amount of combustible matter contained in a given volume. The finer the dust, the more rapid the explosion.

Generally, a dust explosion happens in two stages. First, ignition of the air-dust mixture results in an explosion which in itself not very violent, but is sufficient to agitate dust lying on floors, ledges, etc., to form another dust cloud. The resultant dust cloud is ignited by the flame from the first explosion, and subsequently produces the second explosion - usually of greater intensity.

In order to avoid dust explosions precautions against the formation of dust clouds and against ignition should be taken.

Also see: *Family Tree of Explosions*, and *Successive Explosion.*

Dwell Time – See *Time, Dwell.*

Dynamite - Dynamite is essentially *nitroglycerin* mixed with some absorbent material, giving it a solid form. Dynamite is the first safe high explosive used, enabling man to blast away great masses of rock and other obstacles with comparative safety. It can be dropped, hit with a hammer or even burned with no explosion. Dynamite should have a detonator to set it off.

Alfred Nobel invented dynamite in 1867. It was his second important invention. At the time nitroglycerin was the most powerful explosive in common use, but due to its extreme sensitivity to shock, friction, and heat, it could not be handled with any degree of safety. It proved to be too hazardous and difficult for practical blasting purposes. In the course of his research to find a safe way to carry and use nitroglycerin, Nobel discovered that nitroglycerin was absorbed to dryness by kieselguhr, a porous siliceous earth, and the resulting mixture became much more stable, safer to use and easier to handle. The absorbent substance reduced the sensitivity of nitroglycerin. The new product was named dynamite, from Greek *dynamis*, meaning 'power' by Nobel.

Since its inception basic dynamite has undergone numerous modifications. Instead of kieselguhr, wood pulp, sawdust, charcoal, plaster of Paris, and many other substances came to be used. 'Dope' is also used instead of absorbent.

In the present day, practical context modern dynamite may be defined as a composition explosive essentially comprising, but not limited to, nitroglycerin absorbed in carbonaceous or other fuels (such as wood pulp), oxidizers (such as ammonium nitrate, potassium or sodium nitrate), nitrocellulose, liquid sensitizers (such as nitroglycol), or other similar liquid sensitizers).

Usually ordinary dynamite is packed in sticks of sizes from 2.5 to 5 centimeters (1 to 2 inches) in diameter and about 20 centimeters (8 inches) long. The sticks comprised of brown paper wrappers coated with paraffin to keep out moisture.

In many dynamites the nitroglycerin is diluted with ammonium nitrate. This cools the flame and makes the dynamite safer for use in mines. Most dynamites have the nitroglycerin diluted with another explosive to modify the effect in some way. The use of so-called straight dynamite (nitroglycerin alone) is uncommon.

There are two major classifications within the dynamite family: ***granular dynamite*** and ***gelatin dynamite***. Granular dynamite is a composition that uses nitroglycerin as the explosive base, whereas gelatin dynamite uses a mixture of nitroglycerin and nitrocellulose, which produces a rubbery waterproof composition.

Also see: Entries below for variety of dynamites.

Compare: *ANFO, Slurry, Emulsion, Water Gel, and Cast Booster.*

Dynamite, Ammonia – Ammonia Dynamite: Ammonia dynamite belongs to the *granular dynamite* family.

Ammonia dynamite is a misnomer. It does not contain free ammonia. It is actually dynamite with portion of the liquid sensitizer, such as *nitroglycerin* or *nitroglycol,* and in some cases sodium nitrate, substituted by ammonium nitrate. The dynamite derives the major portion of its energy from reaction of ammonium nitrate.

The replacement by ammonium nitrate makes the explosive slightly less sensitive to shock and friction, and therefore safer to use than *straight dynamite*. It is otherwise similar to *straight dynamite.* In fact, it is *straight dynamite* containing ammonium nitrate. It is the most widely used cartridged high explosive.

Synonym: Extra Dynamite.

Dynamite, Ammonia Gelatin – Ammonia Gelatin Dynamite: It belongs to the *gelatin dynamite* family. This dynamite contains nitroglycerin, nitroglycol, or similar liquid sensitizers and nitrocellulose with a portion of the liquid sensitizer replaced by ammonium nitrate. Ammonia gelatin dynamite is a mixture of straight gelatin with additional ammonium nitrate added to replace some of the nitroglycerin. It is usually used as primers for blasting agents, or as a bottom load in small diameter blast holes.

Synonym: Special or Extra Gelatin.

Compare: *Ammonia Dynamite.*

Dynamite, Ammonium – Synonym of *Ammonia Dynamite.*

Dynamite, Ammonium Gelatin – Ammonium Gelatin Dynamite: Synonym of *Ammonia Gelatin Dynamite.*

Dynamite, Blasting – Blasting Dynamite: It is perhaps the most powerful form of dynamite devised by Alfred Nobel in 1875. It contains nitrocotton colloidally dissolved in nitroglycerin and is waterproof.

Dynamite, Blasting Gelatin – Blasting Gelatin Dynamite: Alfred Nobel invented this dynamite composition. Blasting gelatin is the most powerful industrial explosive. It is the strongest of all nitroglycerine explosives. Alfred Nobel invented this dynamite composition. It is a colloid produced by gelatinizing nitroglycerine with about 7-8% nitrocellulose. *Nitroglycerine* increases the energy of *nitrocellulose* and results in a cleaner burning explosive. This jelly-like material is a translucent, rubber-like elastic and totally unaffected by water.

Varieties of blasting gelatin contain suitable proportions of potassium nitrate and wood-meal.

It has an extremely high velocity and is rated at 100% strength. It is taken as a standard of explosive power. Rate of detonation is 7800 meters (2400 feet) per second (d = 1.63). It is used for special cases of tunnel-driving, shaft-sinking, deep-well shooting, and submarine work.

Blasting gelatin is used as a comparative explosive in determinations of relative weight strength (Ballistic Mortar).

Synonym: Nitroglycerine gelatin, Torpedo Explosive No. 1.

Also see: *Nitroglycerine* and *Cellulose Nitrate.*

Dynamite, Ditching – Nitroglycerin dynamite especially designed and used to propagate sympathetically from hole to hole in ditch blasting.

Also see: *Sympathetic Detonation.*

Dynamite, Extra – Extra Dynamite: Synonym of *Ammonia Dynamite.*

Dynamite, Gelatin – Gelatin Dynamite: It is one of the two major categories of dynamites, the other being granular dynamite. Gelatin dynamite uses a mixture of nitroglycerin (NG) and nitrocellulose (NC) as the explosive base, which produces a rubbery waterproof substance. Ethylene Glycol Dinitrate (EGDN) is usually an additive.

Under the gelatin dynamites classification, there are three sub-classes which are *straight gelatin*, *ammonia gelatin* and *semi gelatin dynamite*.

Compare: *Dynamite, Granular.*

Dynamite, Granular – Granular Dynamite: It is one of the two major categories of dynamites, the other being gelatin dynamite. Granular dynamite uses the compound nitroglycerin as the explosive base.

Under the granular dynamites classification, there are three sub-classes, which are *straight dynamite*, *high-density extra dynamite* and *low-density extra dynamite*.

Compare: *Dynamite,* and *Gelatin.*

Dynamite, Guhr – Guhr Dynamite:

Dynamite, High-density Extra – High-density Extra Dynamite: A sub-class of *granular dynamite*. It is similar to straight dynamite except that some of the nitroglycerin is substituted by ammonium nitrate. The ammonia or extra dynamite is less sensitive to shock and friction than straight dynamite.

This dynamite is the most widely used in mining, quarries and construction applications.

Compare: *Low-density Extra Dynamite.*

Dynamite, Low-density Extra – Low-density Extra Dynamite: A sub-class of *granular dynamite*. It is similar to *high-density extra dynamite* except that more nitroglycerin has been substituted by ammonium nitrate. Its bulk or volume strength is relatively low because the cartridge contains a large proportion of ammonium nitrate.

Low-density extra dynamite is useful in weak rock or where a deliberate effort is made to limit the energy density, the energy per linear length of borehole.

Dynamite, Low-freezing – Low-freezing Dynamite: These are made from nitroglycerin to which additional products have been added. The additional products themselves have low-freezing points and thus lower the freezing point of the mixture.

Synonym: Non-freezing Dynamite.

Dynamite, Low Velocity – See *Low Velocity Dynamite.*

Dynamite, Medium Velocity – See *Medium Velocity Dynamite.*

Dynamite, Nitroglycerin – Nitroglycerin Dynamite: The conventional dynamite. Nitroglycerin dynamite has to be handled with maximum care and caution. It should be kept quiet and away from excessive vibration and contact with metal which serves to sensitize it.

Also see: *Dynamite.*

Dynamite, Non-freezing – Non-freezing Dynamite: Synonym of *Low-freezing Dynamite.*

Dynamite, Non-nitroglycerin – See *Non-nitroglycerin Dynamit.*

Dynamite, Permissible – See *Permissible Dynamite.*

Dynamite, Permissible Ammonia – Permissible Ammonia Dynamite: The Ammonia dynamite to which additions have been made to reduce the amount of toxic fumes evolved, and to lower the temperature of the explosion and reduce the flame.

Also see: *Permissible Dynamite.*

Dynamite, Permissible Gelatin – Permissible Gelatin Dynamite: The gelatin dynamite to which additions have been made to reduce the amount of toxic fumes evolved, and to lower the temperature of the explosion and reduce the flame.

Also see: *Permissible Dynamite.*

Dynamite, Semi-Gelatin – Semi-Gelatin Dynamite: A cross between *ammonia dynamite* and *ammonia gelatin.* Semi-gelatin dynamite is essentially a cross between granular dynamite and gelatin dynamite.

Semi-gelatin dynamite is similar in some respects to ammonium gelatin dynamite except that more of the nitroglycerin/nitrocellulose mixture is substituted by ammonium nitrate. This product is somewhat cohesive and gelatinous. It has more water resistance than granular dynamite because of its gelatinous nature and is more often used under wet conditions. It is sometimes used as a primer for blasting agents.

Dynamite, Straight – Straight Dynamite: The term 'straight dynamite' means that the dynamite contains no ammonium nitrate. Nitroglycerine is the only *high explosive* in straight dynamite.

Straight dynamite is a sub-class of granular dynamite. It is comprised of nitroglycerin, sodium nitrate, carbonaceous fuels (wood pulp and ground shells), sulfur and antacids. Straight dynamite is the most sensitive commercial high explosive in use today. Use in mining or construction applications is avoided because of its sensitivity to shock. Such sensitivity to shock may result in sympathetic detonation in wet blast holes rather than being initiated by the caps within the hole.

However, straight dynamite is an extremely valuable product for blasting ditches because of the fact that in ditching applications normally one detonator is used in the first hole, and subsequent holes are fired by sympathetic detonation. The sympathetic detonation is an attribute in ditching. The sympathetic detonation eliminates the need for a detonator in every hole.

Straight dynamite is not suitable for underground work because of the formation of noxious fumes upon explosion.

Dynamite, Straight Gelatin – Straight Gelatin Dynamite: This product is a sub-class of *gelatin dynamite*. It is straight dynamite in which nitroglycerine has been gelatinized most often with nitrocellulose, and to which nitroglycol, sodium nitrate, carbonaceous fuel (wood pulp and ground shells) and sometimes sulfur are added. Straight gelatin dynamite may also contain ethylene glycol dinitrate (EGDN) and other low molecular weight nitro-compounds.

Straight gelatin dynamite is the most powerful nitroglycerin-based dynamite, and because of its composition, would also be the most water-resistant dynamite.

Earth Electrode System – See *Ground Terminal.*

EBW – Acronym for *Exploding Bridge Wire.*

EBW Cap – See *Cap, EBW.*

EBW Detonator – See *Detonator, EBW.*

EC Blank Fire – Synonym of *EC Smokeless Powder.*

EC Powder – Acronym for Explosive Company (EC) Powder. It is a type of modified nitrocellulose powder used for fragmentation charges in grenades and the like. It is coarser than regular smokeless powder. EC Powder has additional nitrates added to it.

EC Smokeless Powder – EC Smokeless Powder is used as a charge in small arms in blank cartridges. The explosive powder resembles course sand. Its color is orange or pink.

Synonyms: *Blank-fire Powder*, and *EC Blank Fire.*

EDD – Acronym for *Electric Delay Detonator.*

Edible Explosive – Trinitroglycerin, the active ingredient of dynamite, is also used medically as a vasodilator. A very useful heart drug is made up of trinitroglycerin. The low concentration of nitroglycerine in medication, and the moist environment in the body, make it impossible for it to explode. The nitroglycerine that is absorbed by the body dilates the blood vessels, increases the blood supply to the heart, and reduces the workload of sending blood around a body. Angina pain is relieved as the heart muscle receives the oxygenated blood it needs for good functioning.

EED – Acronym for Electro Explosive Device. An EED is a device containing some reaction mixture of explosive that is initiated by an electric current. It is in effect a detonator or initiator that is initiated by an electric current. Cartridge, squib, igniter, etc., which is electrically initiated are examples of EED. The output of such initiation is heat, shock, or mechanical action.

Also see: *Low-energy EED.*

EED, Low-energy – See *Low-energy EED.*

EFF – Acronym for Explosive Forged Fragment.

Efficient Artificial Barricade – It means an artificial mound or properly riveted wall of earth, around a storage magazine, the minimum thickness of which is not less than one meter (three feet).

EFP – Acronym for both Explosive Formed Projectile and Explosive Forged Projectile.

EGDN – Abbreviation of *Ethylene Glycol Dinitrate.*

Electrical Bonding – Electrical connection between two conductive objects intended to prevent development of an electrical potential between them.

Electric Blasting – See *Blasting, Electric.*

Electric Blasting Circuit – See *Blasting Circuit, Electric.*

Electric Cap – See *Cap, Electric.*

Electric Cap Lamp – See *Lamp, Electric Cap.*

Electric Coal Drill – See *Coal Drill, Electric.*

Electric Delay Blasting Cap – See *Detonator, Electric Delay*.

Electric Delay Detonator – See *Detonator, Electric Delay*.

Electric Detonator – See *Detonator, Electric.*

Electric Fuse – See *Fuse, Electric.*

Electric Igniter – See *Igniter, Electric.*

Electric Match – A device designed and used for the electrical ignition of fireworks and pyrotechnic articles that contains a small amount of pyrotechnic material, which ignites when a specified electric current flows through the leads.

Synonym: *Igniter.*

Electric Primer – A device made of metal, which contains a small amount of sensitive explosive or charge of black powder and is actuated by energizing an electric circuit. It is designed and used to set off explosive or propelling charges.

Electric Squib – See *Squib, Electric.*

Electro Explosive Device – See *EED.*

Electronic Detonator – See *Detonator, Electronic.*

Empty Cartridge Bag – Black Powder Igniter – The phrase means empty bags attached to an igniter composed of black powder.

Empty Cartridge Shell, Primed – See *Primed Empty Cartridge Shell.*

Empty Grenade, Primed – See *Primed Empty Grenade.*

Emulsion Explosives – An important type of explosives used in commercial blasting. A category of blasting agent. Emulsion explosives are comprised of droplets of an immiscible fuel surrounded by water containing substantial amount of oxidizer (oil in water emulsions), or substantial amount of oxidizer dissolved in water droplets surrounded by an immiscible fuel (water in oil emulsions), to which sensitizers and emulsifying agents are usually added. Emulsion explosives are intimate and homogeneous mixtures of oxidizers and fuels.

In this water-in-oil emulsion, microscopically fine droplets of aqueous solution of oxidizer such as ammonium-, calcium- or sodium nitrate is finely dispersed into the continuous phase of fuel oil. An emulsifying agent such as sodium oblate or sodium mono-oblate is added to prevent liquid separation. Sensitizers such as perchlorates salts are added which also help to improve shelf life. Dispersed gas can be put into the emulsion matrix in order to control density within a range 0.70 to 1.35 gm/cm^3. This is achieved with micro balloons or by chemical gassing of the composition.

Emulsion explosives are similar to slurry in some respects.

Since each constituent of this emulsion is coated with an oily film, the emulsions have excellent water resistance. The high strength and *VOD* of emulsion makes it an ideal product that matches the performance of gelatins without the headaches.

Emulsion explosives are used for blasting in quarries and stripping in underground metal mines, tunneling, etc.

Also see: *Dynamite*, *ANFO*, *Slurry*, *Water Gel*, and *Cast Booster*.

End Burning – See *Burning, End.*

End Spit – The flash of burning material ejected from safety fuse when burning reaches a cut end.

Energized Explosive – An explosive composition to which another fuel having a high heat of combustion, such as aluminum, beryllium, magnesium, sulfur, or zinc, is added to increase the heat of explosion.

Also see: *Aluminized Explosive.*

Energy – A measure of the potential for the explosive to perform work.

Energy, Bubble – See *Bubble Energy.*

Energy, Shock – See *Shock Energy.*

Entire Load – Synonym of *Total Contents.*

EOD – Acronym for *Explosive Ordnance Disposal.*

Epicenter – In relation to an *unconfined vapor cloud explosion*, the ground location beneath the deduced center of the explosion.

Eq.S – Acronym for Equivalent to Sheathed. The original name for Group P3 permitted explosives.

Ethylene Dinitrate – Synonym of Ethylene Glycol Dinitrate, and *Nitroglycol.*

Ethylene Glycol Dinitrate – Synonym of *Nitroglycol.*

Ethyne – Synonym of *Acetylene.*

Excitation Time – See *Time, Excitation.*

Expelling Charge – A quantity of propellant used in a special purpose shell designed to eject the contents of the shell without damage.

Explode – The term 'explode' is generally used to mean explosive effects capable of endangering life and property through blast, fragments and missiles.

Synonyms: blow up, break loose, burst, burst forth, set off.

Antonym: *Implode.*

Exploder – A device for firing or detonating an explosive charge as blasting cap, blasting machine or squib.

Exploding, Composition – See *Composition Exploding.*

Exploding Bridge Wire – See *EBW*, and *EBW Detonator.*

Exploding Bridge Wire Detonator – *See EBW,* and *Detonator, EBW.*

Explosibility – The quality of being explosible, or the capacity for being exploded.

Explosible – Capable of being exploded.

Explosimeter – An instrument for testing explosibility by measuring the concentration of combustible gases and vapors in air.

Explosion – Explosion means an act of bursting or exploding. It is a violent expansion or bursting that is caused by a sudden release of energy, from a very rapid chemical reaction, from a nuclear reaction, or from an escape of gases under pressure as in a gas cylinder, which is accompanied by loud noise.

Thus, an explosion may be defined generally as a phenomenon of the sudden and noisy release of the previously confined energy, affecting the surroundings. That means an explosion is a sudden and rapid conversion of potential energy into kinetic energy, with the release or production and release of gases under pressure accompanied by heat, shock and a loud noise.

An explosion is effected by:

- a very fast burning reaction;
- detonating an explosive;
- bursting a vessel containing a pressurized fluid;
- rapid heating of air and plasma by an electric arc; or
- a nuclear reaction.

Harmless bursting of a toy balloon or disastrous nuclear reaction of an atom bomb fall under the definition of explosions. However, the use of the term 'explosion' is usually confined to describing incidents in which the rapid release or production and release of energy causes a significant blast wave capable of causing damage.

In an explosion, the release or production and release of gases under pressure accompanied by heat, shock and loud noise causes a significant blast wave capable of causing damage to surrounding property. In some cases, the resultant high-pressure gases of an explosion are utilized to perform physical work. Some explosions are desirable and controlled, as in blasting in quarries and mines or in the cylinder of an internal combustion engine.

An explosion may occur without an *explosive*. A steam boiler may explode; the heat energy which causes the explosion is not intrinsic to water, it (heat energy) is put into the water in the boiler. Hence water is not an explosive, and no chemical reaction takes place in this physical explosion.

There are a number of ways of categorizing explosions, and any given explosion incident may fall into more than one category. Basically explosions can be divided into three categories:

Physical,
Chemical, and
Nuclear.

Antonym: *Implosion.*

Also see: *Family Tree of Explosions*; and individual entries hereunder.

Explosion, Aerobic – See *Aerobic Explosion.*

Explosion, Aerosol – See *Aerosol Explosion.*

Explosion, Anaerobic – See *Anaerobic Explosion.*

Explosion, Atomic – Atomic Explosion: See *Nuclear Explosion.*

Explosion Bonding – See *Welding, Explosion.*

Explosion, Chemical – See *Chemical Explosion.*

Explosion Cladding – See *Welding, Explosion.*

Explosion, Combustion – See *Combustion Explosion.*

Explosion, Confined – See *Confined Explosion.*

Explosion, Controlled – See *Controlled Explosion.*

Explosion, Decomposition – See *Decomposition Explosion.*

Explosion, Dense – See *Dense Explosion.*

Explosion, Deflagration – See *Deflagration Explosion.*

Explosion, Desired – See *Desired Explosion.*

Explosion, Detonation – See *Detonation Explosion.*

Explosion, Dust – See *Dust Explosion.*

Explosion Efficiency – The term 'Explosion Efficiency' means the ratio of the energy in the blast wave to the energy theoretically available from the heat of combustion, usually expressed as a percentage.

Explosion, Family Tree of – See *Family Tree of Explosion.*

Explosion, Flareless – Flareless Explosion: The term means an explosion without a flame.

Explosion, Gas – See *Gas Explosion.*

Explosion Hazard – The damage that arises from an explosion may be caused either by the effect of the blast or shock wave, or by missiles. Part of the energy liberated in an explosion may be imparted to fragments or whole systems in the form of kinetic energy. These fragments, or missiles, may be projected outwards some considerable distance from the center of an explosion. In a chemical explosion large volumes of highly heated gases are produced that exert enormous pressures in the surrounding medium. That such products of reaction are produced in such a rapid fashion, milliseconds or less, provides opportunity for much useful work and makes them essential tools for mining, quarrying or construction industries. However, the same characteristics present risks to persons working with explosives as well as installations surrounding the center of the explosion if it happens at an unexpected location, unintended time or in an unplanned manner. The useful work sought from such explosives may become a devastating potential for tragic loss.

Explosion Heat – Synonym of *Heat of Explosion.*

Explosion, Heat of – See *Heat of Explosion.*

Explosion, Heterogeneous – See *Heterogeneous explosion.*

Explosion, Homogeneous – See *Homogeneous Explosion.*

Explosion, Mass – Mass Explosion: A mass explosion is one that affects the entire load almost instantaneously.

Explosion, Mechanical – Mechanical Explosion: Synonym of *Physical Explosion.*

Explosion, Non-Seated – See *Non-Seated Explosion.*

Explosion, Nuclear – See *Nuclear Explosion.*

Explosion, Nuclear Fission – See *Nuclear Fission Explosion.*

Explosion, Nuclear Fusion – See *Nuclear Fusion Explosion.*

Explosion of the Total Content – The term ‘explosion of the total content’ is used in testing such a substantial proportion that the practical hazard should be assessed by assuming simultaneous explosion of the whole of the explosive.

Explosion, Open Air – See *Open Air Explosion.*

Explosion, Particulate – See *Particulate Explosion.*

Explosion, Photochemical – See *Photochemical Explosion.*

Explosion, Physical – See *Physical Explosion.*

Explosion Pressure – Synonym of *Borehole Pressure.*

Explosion Proof – In short, explosion proof means an apparatus that contains an explosion inside an enclosure. The apparatus is enclosed in a case which is capable of withstanding an explosion of a specified gas or vapor that may occur within it. It can prevent the ignition of a specified gas or vapor surrounding the enclosure (by sparks or flashes), and operate at such an external temperature that a surrounding flammable atmosphere will not be ignited.

Explosion, Propagation of – Propagation of Explosion: See *Sympathetic Explosion.*

Explosion Range – In relation to gas or vapor-air mixtures and in dust-air mixtures, the region between the *LEL* and the *UEL* is called the explosion range, within which an ignition can cause a self-propagation reaction.

Also see: *LEL* and *UEL*

Explosion, Rarefied – See *Rarefied Explosion.*

Explosion, Salt Bath – Salt Bath Explosion: Salt baths run at high temperatures, and while the bath itself is not flammable, it can promptly ignite combustibles that come in contact with it. Introduction of water into high temperature baths may cause steam or salt eruptions to occur. Mixing unlike salts carries the danger of potential explosion. Such phenomenon is referred to as salt bath explosion.

Explosion, Seated – See *Seated Explosion.*

Explosion, Spontaneous – See *Spontaneous Explosion.*

Explosion, Successive – See *Successive Explosion.*

Explosion Suppression – The technique of detecting and arresting combustion in a confined space while the combustion is still in its incipient stage, thus preventing the development of pressures that could result in an explosion.

Explosion, Sympathetic – Sympathetic Explosion: The explosion of an explosive material by the detonation of another charge in the vicinity without actual contact. A sympathetic explosion may be termed as the initiation of an explosive charge without a priming device as a result of receiving an impulse from another explosion in the neighborhood through air, earth, or water. The second explosion, known as the sympathetic one, is initiated by influence.

The distance through which transmission of detonation may occur depends on many factors, such as mass, physical and chemical characteristics, detonation properties of donor charge, existence and characteristics of acceptor charge confinement, existence and characteristics of the medium between the charge, etc. Other relevant factors being constant, the possibility of the propagation of detonation is determined mostly by both the sensitivity of the acceptor charge and by the initiating strength of the donor charge. Possible factors that contribute to sympathetic detonation also include the initiation of acceptor charge by the passage of shock wave from one mass to the other, by flying fragments and the like.

Synonyms: Sympathetic Detonation, and Propagation of Explosion.

Explosion, Thermal – See *Thermal Explosion.*

Explosion, Transmission of – Transmission of Explosion: Synonym of *Sympathetic Detonation.*

Explosion Triangle – A chemical explosion occurs when a fuel and oxidizer undergoes chemical decomposition or a reaction takes place between fuel and oxidizer on ignition. That means three things, namely fuel, oxidizer and ignition are essential for a chemical reaction to occur which can be represented by the three arms of a triangle.

Fuel +Oxidizer+ Ignition = Explosion.
Synonym: Explosion Triangle.

Fuel
Ignition
Oxidizer

Compare: *Fire Triangle.*

Explosion, Unconfined – See *Unconfined Explosion.*

Explosion, Unconfined Vapor Cloud – See *Unconfined Vapor Cloud Explosion.*

Explosion, Uncontrolled – Uncontrolled Explosion.

Explosion, Undesired – Undesired Explosion.

Explosion, Vapor – See *Vapor Explosion.*

Explosion, Vapor Cloud – See *Vapor Cloud Explosion.*

Explosion, Velocity of – See *Velocity of Explosion.*

Explosion Welding – See *Welding, Explosion.*

Explosive – An intrinsically energetic solid or liquid chemical compound, mixture, or device, which, on being heated, struck or detonated, is capable by chemical reaction of producing an explosion by its own energy without the participation of external reactants such as atmospheric oxygen, with substantially instantaneous production of heat and gas.

In general, an explosive has three basic characteristics:

(1) It is an intrinsically energetic unstable chemical compound or mixture ignited by detonation, friction, heat, impact, shock, or a combination of these conditions;
(2) Upon ignition it decomposes very rapidly in a detonation (as opposed to a deflagration, which is a slower decomposition as with ignition of gunpowder); and
(3) Upon detonation there is a rapid release of heat and large quantities of high-pressure gases, which expand rapidly with sufficient force to overcome confining forces, such as the confining forces of surrounding rock formation.

An explosion may occur without an *explosive.* A steam boiler may explode; the heat energy which causes the explosion is not intrinsic to water, it (heat energy) is put into the water in the boiler. Hence water is not an explosive, and no chemical reaction takes place in this physical explosion.

Although there are many common substances that can produce respectable explosions such as gasoline, these types of materials are not usually classed as explosives because they must obtain their oxidizer from the air. An explosive is an intrinsically energetic unstable chemical substances where generally the fuel and oxidizer are combined in the same material, and which takes part in a chemical explosion. Although wood has higher energy content than most explosives, wood is not an explosive because its energy of activation is much greater, and its energy is released over a much longer period.

For a substance to be classed as an explosive it must decompose or rearrange itself very rapidly upon the application of an initiating force such as detonation, heat or shock, and its combustion must produce heat and gases.

The term 'explosive' applies to materials that undergo either detonation or deflagration. Some explosives must be physically confined to obtain an explosion, while others do not.

An explosive does not occur naturally but is entirely man-made.

Cuprous acetylide is an explosive that explodes by decomposing into copper and carbon and heat; no gas is generated but the sudden heat causes a sudden expansion of the air in the neighborhood, and the result is an unequivocal explosion. Pyrotechnic substances are explosives even when they do not evolve gases.

The term 'Explosive' represents a broad class of inherently dangerous materials, including blasting explosives, fireworks, propellants, ammunition and other explosive devices. The term includes, but is not limited to, black powder, nitroglycerin, dynamite, gunpowder, propellant, detonators, safety fuses, igniters, squibs, and detonating cord.

Also see: The entries hereunder.

Antonyms: Non-explosive.

Explosive-Actuated Device – Explosive-actuated device means any special mechanized device or tool that is actuated by explosives, such as jet perforators or jet-tappers. Propellant actuated devices are not included in this term.

Also see: *Propellant-Actuated Device.*

Explosive, Aerobic – See *Aerobic Explosive.*

Explosive, Aluminized – See *Aluminized Explosive.*

Explosive, Ammonium Nitrate – See *Ammonium Nitrate Explosive.*

Explosive, Ammonium Perchlorate – See *Ammonium Perchlorate Explosive.*

Explosive, Anaerobic – See *Anaerobic Explosive.*

Explosive, Application of – See *Application of Explosive.*

Explosive Article – Synonym of *Explosive Device.*

Explosive, Authorized – See *Authorized Explosive.*

Explosive, Auxiliary – See *Auxiliary Explosive.*

Explosive Belt – A vest packed with explosives and armed with a detonator, worn by suicide bombers. An explosive belt is usually packed with bolts, nails, screws, and other objects that serve as shrapnel to maximize the number of casualties in the explosion.

Synonym: Suicide Belt, and Suicide Vest.

Explosive, Binary – See *Binary Explosive.*

Explosive Bolt – A bolt which is intended to be fractured at a predetermined point by a contained or inserted explosive charge for the purpose of releasing a load.

Explosive Bomb – Any explosive or incendiary material designed and constructed which when dropped, placed, projected or thrown, and initiated in any particular manner, will produce a violent release of high pressure and/or heat.

Explosive Bonding – See Welding, Explosive.

Explosive, Booster – See *Booster Explosive.*

Explosive Bullet – A bullet containing a substance that is exploded by percussion.

Synonym: Percussion Bullet.

Explosive, Bursting – See *Bursting Explosive.*

Explosive, Bursting-Charge – See *Bursting-Charge Explosive.*

Explosive, Cap Sensitive – See *Cap-Sensitive Explosive.*

Explosive Charge – It is the quantity of explosive to be set off at one time; or used in an explosive device, in an industrial application to produce a specific effect such as in a blast hole, coyote tunnel, or other form of placement.

Also see: *Bursting Charge*, *Expelling Charge*, and *Propelling Charge*.

Explosive Charge, Commercial – See *Commercial Explosive Charge*.

Explosive Charge, Main – See *Main Explosive Charge*.

Explosive Charge, Supplementary – See *Supplementary Explosive Charge.*

Explosive, Chemical Constituent of – See *Chemical Constituent of Explosive.*

Explosive Cladding – See *Welding, Explosion*

Explosive, Commercial – See *Commercial Explosive*.

Explosive Company Powder – See *Powder, EC.*

Explosive, Composite – See *Composite Explosive.*

Explosive, Composition – See *Composition Explosive*.

Explosive Compound, Aliphatic – See *Aliphatic Explosive Compound.*

Explosive Compound, Aromatic – See *Aromatic Explosive Compound.*

Explosive, Cool – Cool Explosive: See *Nitroguanidine,* the synonym of cool explosive.

Explosive Cord – One of the synonyms of *Detonating Cord.*

Explosive D – See *Ammonium Picrate*, the synonym of Explosive D.

Explosive Decontamination – The removal of hazardous explosive material.

Explosive, Deflagrating – Deflagrating Explosive: A deflagrating explosive is one that reacts by deflagration rather than by detonation when used in its normal manner. Propellants belong to this type.

Explosive Deflagration – A *deflagration* in which a blast wave is produced having the potential to cause serious damage.

Also see: *Deflagration.*

Explosive, Detonating – A detonating explosive reacts by detonation rather than by deflagration when used in its normal manner.

Explosive, Detonative – Detonative Explosive: Explosive having the property of detonating or characterized by detonation. Detonative (adjective) means exploding almost instantaneously.

Explosive, Detonator Insensitive – See *Detonator Insensitive Explosive*: Synonym of *Blasting Agent.*

Antonym: *Detonator Sensitive Explosive.*

Explosive, Detonator Sensitive – See *Detonator Sensitive Explosive.*

Explosive Device – A device that bursts with sudden violence from internal energy.

Also see: Bomb, Cap, Detonating Device, Detonator.

Explosive Dispersion Device – See *Dirty Bomb.*

Explosive, Disrupting – See *Disrupting Explosive.*

Explosive, Dual Use – See *Dual Use Explosive.*

Explosive, Edible – See *Edible Explosive***.**

Explosive, Emulsion – See *Emulsion Explosive.*

Explosive, Energized – See *Energized Explosive.*

Explosive Energy – The thermochemical heat of reaction of an explosive. Today explosive energy for a given product is generally described in one of four ways:

- Absolute Weight Strength (AWS): The heat of reaction available in each gram of explosive.

- Absolute Bulk Strength (ABS): The heat reaction available in each cubic centimeter of explosive.

- Relative Weight Strength (RWS): The heat of reaction per unit weight of an explosive compared to the energy of an equal weight of standard ANFO.

- Relative Bulk Strength (RBS): The heat of reaction per unit volume of an explosive compared to the energy of an equal volume of standard ANFO at a given density.

Explosive Entry – The utilization of explosive devices to facilitate access into a target area through a conventional or non-conventional breach point.

Explosive Expelling Charge for fire extinguisher – See Charge for fire extinguisher, Explosive Expelling.

Explosive Facility – See *Facility, Explosive.*

Explosive, Flashless – Flashless Explosive: Synonym of *Nitroguanidine*,

Explosive, Forbidden – See *Forbidden Explosive.*

Explosive, Fuel-Air – See *Fuel Air Explosive.*

Explosive Hazard Classification – See *Hazard Classification, Explosive.*

Explosive, Her/His Majesty's – See *Her/His Majesty's Explosive.*

Explosive, High – See *High Explosive.*

Explosive, High, Insensitive – See *Insensitive High Explosive.*

Explosive, High Melting – See *High Melting Explosive.*

Explosive, High, Primary – See *Primary High Explosive.*

Explosive, High, Secondary – See *Secondary High Explosive.*

Explosive, High, Tertiary – See *Tertiary High Explosive.*

Explosive, Hydrated – See *Hydrated Explosive***.**

Explosive, Impulse – See *Impulse Explosive.*

Explosive, Initiating – See *Initiating Explosive.*

Synonym: *Primary Explosive.*

Explosive, Inorganic – See *Inorganic Explosive.*

Explosive, Insensitive, Detonator – See Detonator Insensitive Explosive.

Synonym: *Blasting Agent.*

Explosive, Insensitive High – See *Insensitive High Explosive.*

Explosive, Intermediate – See *Intermediate Explosives.*

Explosive Lead – See *Lead***.**

Explosive Limit – The upper and lower limits of percentage composition of a combustible gas mixed with other gases or air within which the mixture explodes when ignited.

Explosive Limit, Lower – See *Lower Explosive Limit (UEL).*

Explosive Limit, Upper – See *Upper Explosive Limit (UEL).*

Explosive, Liquid Carbon Dioxide – See *Liquid Carbon Dioxide Explosive.*

Explosive, Liquid Oxygen – See *Liquid Oxygen Explosive.*

Explosive Loading Factor – Explosive loading factor means the amount of explosive used per unit of rock. Also called powder factor.

Synonym: *Powder Factor.*

Explosive, Low – See *Low Explosive.*

Explosive Material – The term 'explosive material' includes, but is not necessarily limited to, black powder, blasting agent, dynamite and other high explosives, emulsion, slurry, water gel, pellet powder, initiating explosives, detonators, safety fuses, squibs, detonating cord, igniter cord, and igniters.

An explosive material has the following characteristics:

- It is chemically or otherwise energetically unstable.
- The initiation produces a sudden expansion of the material accompanied by large changes in pressure (and typically a flash or loud noise), which is the explosion.

Explosive Material, Bullet-sensitive – See *Bullet-sensitive Explosive Material.*

Explosive, Medicinal – See *Medicinal Explosive.*

Explosive, Metallized – See *Aluminized Explosive* and *Metallized Explosive.*

Explosive Mine – Explosive mine means metal containers filled with a high explosive and provided with a detonating device.

Explosive Mixture – See *Mixture, Explosives.*

Explosive, Mock – See *Mock Explosive.*

Explosive Motor – Explosive Motor is a device in which explosive is completely enclosed and which upon operation causes a mechanical movement as of a piston.

Explosive, Nitrate Mixture – See *Nitrate Mixture Explosive.*

Explosive, Nitramine – See *Nitramine Explosive.*

Explosive, Nitro compound – See *Nitro Compound Explosive.*

Explosive No. 1, Torpedo – See Torpedo Explosive No. 1.

Explosive, Non-cap Sensitive – See Non-cap Sensitive Explosive.

Explosive, Nuclear – See *Nuclear Explosive.*

Explosive Oil – See *Oil, Explosive*.

Explosive, on Ignition, Behavior of – See *Behavior of Explosive on Ignition.*

Explosive Operation, Nuclear – See *Nuclear Explosive Operation.*

Explosive Ordnance Disposal – The term is used to describe military operations in which hazardous devices are rendered safe.

Acronym: *EOD.*

Explosive, Peaceful Uses of – See *Uses of Explosive.*

Explosive, Permissible – See *Permissible Explosive.*

Explosive, Permitted – See *Permitted Explosive.*

Explosive, Picatinny Liquid – See *Picatinny Liquid Explosive (PLX).*

Explosive, Plastic – See *Plastic Explosive.*

Explosive, Polymer Bonded – See *Polymer Bonded Explosive.*

Explosive Power Device – An explosive device designed and used to accomplish mechanical actions. It consists of housing with a charge of deflagrating explosive and a means of ignition. The gaseous products of the deflagration produce linear or rotary motion or function diaphragms, valves or switches.

Synonym: *Power Device Cartridge.*

Explosive, Primary – See *Primary Explosive.*

Explosive, Primary High – See *Primary High Explosive.*

Explosive Projectile – See *Projectile, Explosive.*

Explosive, Propellant – See *Propellant Explosive.*

Explosive Range – See *Range, Explosive.*

Explosive Reaction – In most explosives (but not all), the process by which large quantities of heat and gas are produced is oxidation, i.e. combustion in which the oxidizer is provided by the explosive, unlike normal combustion where the oxidizer is atmospheric oxygen. Once this oxidation process in the explosive has been initiated, it will proceed without any additional energy or material being required, until the explosive has been fully consumed. In some explosives, such as black powder, the fuels (carbon and sulphur) are present as separate ingredients mixed intimately with the oxidizer (potassium nitrate). In other explosives, the fuel and oxidizer are present within the same molecule, as in trinitrotoluene (TNT) where the fuels are carbon and hydrogen atoms and the oxidizers are nitro (NO_2) groups.

Explosive Reaction, Initiation of – See *Initiation of Explosive Reaction.*

Explosive Relay – See *Relay, Explosive.*

Explosive Release Device – See *Release Device, Explosive.*

Explosive Rivet – See *Rivet, Explosive.*

Explosive, Safety – See *Safety Explosive.*

Explosive, Secondary – See *Secondary Explosive.*

Explosive, Secondary High – See *Secondary High Explosive.*

Explosive, Seismic – See *Seismic Explosive.*

Explosive, Self-contained – Self-contained Explosive: The organic nitrates (nitrocellulose, nitroglycerine, nitrostarch) contain the oxygen and the oxidizable elements carbon and hydrogen within the same molecule; they are self-contained explosives. The same is true for the aromatic nitro explosives (trinitrotoluene, trinitrophenol).

Also see: *Explosophore Groups.*

Explosives Engineer – *An Explosives Engineer is responsible for breaking rock to make way for houses, buildings, roads and bridges as well as extracting minerals, metals, and fuels from the ground to provide power, raw materials, and products for manufacturing and consumer goods. [*Quoted from ISEE publications.]

Explosives Engineering – *Explosives Engineering is defined as that area of engineering practice where judgment and experience are utilized in the application of scientific principles and techniques as they pertain to the use, handling manufacture, transportation and storage of explosives. [*Quoted from ISEE publications.]

Explosive, Detonator Sensitive – See *Detonator Sensitive Explosive.*

Explosive, Sheathed – See *Sheathed explosive.*

Explosive, Sheet – See *Sheet Explosive.*

Explosive Shell, High – See *High Explosive Shell.*

Explosive Sign – The sign placed near a storage area as a warning to unauthorized personnel, or on a vehicle transporting explosives denoting the character of the cargo. Usually such a sign is called placard.

Explosive, Slurry – See *Slurry Explosive.*

Explosive Sounding Device – Devices containing an explosive charge and a means of initiation. They function when, after having dropped from a ship, they hit the seabed.

Explosive, Sprengel – See *Sprengel Explosive.*

Explosive Storage Compatibility Group (SCG) – Under the UNO system, there are 13 storage compatibility groupings, which further categorize Class 1 explosives by their form or composition, ease of ignition, and sensitivity to detonation. Each Storage Compatibility Group (SCG) is described below:

SCG A – Primary explosive substance.

SCG B – Articles containing a primary explosive substance and not containing two or more independent safety features.
Examples are detonators, small arms primers, fuses, and detonators of all types.

SCG C – Propellant explosive substance, other deflagrating explosive substance, or article containing such substance. Examples are single-, double-, and triple-base propellants; composite propellants; rocket motors (solid propellant); and ammunition with inert projectiles.

SCG D – Secondary detonating explosive substance or black powder or article containing a secondary detonating explosive, in each case without its own means of initiation and without a propelling charge, or article containing a primary explosive substance and containing two or more independent safety features.

SCG E – Article containing a secondary detonating explosive substance, without means of initiation, with a propelling charge (other than one containing a flammable or hypergolic liquid). Examples are artillery ammunition, rockets, and guided missiles.

SCG F – Article containing a secondary detonating explosive substance with its own means of initiation, with a propelling charge (other than one containing a flammable or hypergolic liquid) or without a propelling charge. Examples are grenades, sounding devices, and similar items with an in-line explosive train in the initiator.

SCG G – Pyrotechnic materials and devices containing pyrotechnic materials. Examples are devices that, when functioning, result in illumination, smoke, or an incendiary, lachrymatory, or sound effect.

SCG H – Article containing both explosive substance and white phosphorus.

SCG J – Article (explosive device) containing both explosives and flammable liquids or gels. Examples are liquid- or gel-filled incendiary ammunition, fuel-air explosive (FAE) devices, flammable liquid-fueled missiles, and torpedoes.

SCG K – Article containing both explosives and toxic chemical agents. Examples are artillery or mortar ammunition (fused or unfused), grenades, and rockets or bombs filled with a lethal or incapacitating chemical agent.

SCG L – Explosive substance or article containing an explosive substance and presenting a special risk needing isolation of each type. Examples are damaged or suspect explosives devices or containers, explosives that have undergone severe testing, fuel-air explosive devices, and water-activated devices.

SCG N – Hazard Division 1.6 ammunition containing only extremely insensitive detonating substances. Examples are bombs and warheads.

SCG S – Explosive substance or article so packed or designated that any hazardous effects are confined and self-contained within the item or package. Examples are thermal batteries, cable cutters, explosive actuators, and other ammunition items packaged to meet the criteria of this group.

Explosive Strength – See *Strength, Explosive.*

Explosive Substance – 1. An explosive substance is a single molecule substance (such as TNT), containing a molecular group that possesses explosive properties (such as azides). Explosive Substances are a sub-class of chemical explosives, the other being explosive mixtures (composite explosives). It means the chemical proper, either a compound or a mixture, not being enclosed in any case or contrivance to make it ammunition or a device.

Also see: *Composite Explosive* and *Explosive Article.*

Explosive Switch – A self-contained electrically initiated small unit that causes one or more electric circuits to be opened and/or closed by a propulsive explosive action.

Synonym: *Squib Switch.*

Explosive, Tertiary – See *Tertiary Explosive.*

Explosive, Tertiary High – See *Tertiary High Explosive.*

Explosive Torpedo – A metal device containing a means of propulsion and a quantity of high explosives.

Explosive Train – See *Train, Explosive.*

Synonyms: Firing Train, and Initiation Sequence.

Explosive Train Component – See *Train Component, Explosive.*

Explosive, Two-component – See *Two-component Explosive.*

Explosive Unit – A unit for the measurement of the force of an explosion.

Also see: *Kiloton*, *Megaton*, and *Unit of Measurement.*

Explosive, Use of – See *Use of Explosive.*

Explosive Wave – A wave of chemical action which passes through an explosive substance when it explodes; also, more accurately, detonation zone.

Explosive Weapon – See *Weapon, Explosive.*

Explosive Welding – See *Welding, Explosive.*

Explosophore Group – With the exception of a few azides and acetylides, which find their use as initiators, most explosives contain oxygen and oxidizable elements. Oxygen is usually present in the form of labile radicals, such as the nitric acid radical $-ONO_2$ or the nitro group $-NO_2$. Such groups are characteristic for explosives, and they are called explosophore groups. Other oxygen-containing explosophore groups are the chlorate radical $-ClO_3$, the perchlorate radical $-ClO_4$, the fulminate radical -ONC, and the nitramine group $-NHNO_2$. The inorganic nitrates (potassium, sodium, ammoium, barium) are used as components of explosive mixtures in combination with oxidizable materials such as charcoal, sulfur in elementary or combined form, nitro compounds having insufficient oxygen for complete combustion, metals, metal alloys (such as silicides), etc. The organic nitrates (nitrocellulose, nitroglycerine, nitrostarch) contain the oxygen and the oxidizable elements carbon and hydrogen within the same molecule; they are self-contained explosives. The same is true for the aromatic nitro explosives (trinitrotoluene, trinitrophenol). Some explosives, like nitroglycerine and ammonium nitrate, have an excess of oxygen; others, like trinitrotoluene, do not have a sufficient amount of oxygen for oxidation of the oxidizable elements present. A negative oxygen balance can be corrected by addition of substances that have an excess of oxygen, such as ammonium nitrate.

Extra Dynamite – Synonym of *Ammonia Dynamite.*

Extra Dynamite, Low-density – See *Low-density Extra Dynamite.*

Extra Gelatin Dynamite – Synonym of *Ammonia Gelatin Dynamite.*

Extraneous Electricity – Electrical energy, other than actual firing current or the test current from a blasting galvanometer, that is present at a blast site and that could enter an electric blasting circuit. It includes static electricity, stray current, lightning, radio-frequency waves, and time-varying electric and magnetic fields. Extraneous electricity may be a hazard with electric detonators.

Extrusion – It is one of the three methods of loading high explosive charges into their containers, the other two methods being cast-loaded and pre-loaded. The combining of certain explosives results in plastic mixtures that can be loaded only by the extrusion method. The extrusion method employs a pressure system for forcing the plastic mixtures into the various types of projectiles and bomb casings.

Exudation – The separation of oily ingredients out of explosives during prolonged storage. In nitroglycerine gelatin explosives, exudation means liberation of nitroglycerine following breakdown of the gelatinous base. If the percentage of nitroglycerine, aromatic compounds or gelatinizers is high in propellant charges, there remains a possibility of exudation.

Exudation is also known to occur in the case of cast TNT. This exudate may appear pale yellow to brown and may vary in consistency from an oily liquid to a sticky substance. The amount and rate of separation depend upon the purity of the TNT and upon the temperature of the stowage place. Low-melting point (Grade B) TNT may exude considerable liquid and generate some gas. The exudation is accelerated with an increase in temperature.

Exudation of unbonded nitroglycerine is highly dangerous. Accumulation of exudate is a great risk of explosion and fire. Its accumulation should always be avoided by continual removal and disposal as it occurs.

Proper gelatinizing of nitroglycerine by the addition of a small percentage of nitrocellulose assists in preventing exudation, or settling out of the absorbent material. Turning over the boxes of stored dynamites at regular intervals will reverse the settling flow.

Synonym: *Weeping*.

Extruded Booster – See *Booster, Extruded*.

Facility – A group of buildings or equipment used for explosive operations at one geographic location.

Facility, Explosive – Explosive Facility: A structure or defined area used for explosive storage or operations.

FAE – Acronym for *Fuel Air Explosive*. See Vacuum Bomb, the synonym of FAE.

Fall Hammer Test – The test intended and designed to measure the sensitiveness of an explosive to impact using a weight that falls vertically.

Family Tree of Explosion – See next page.

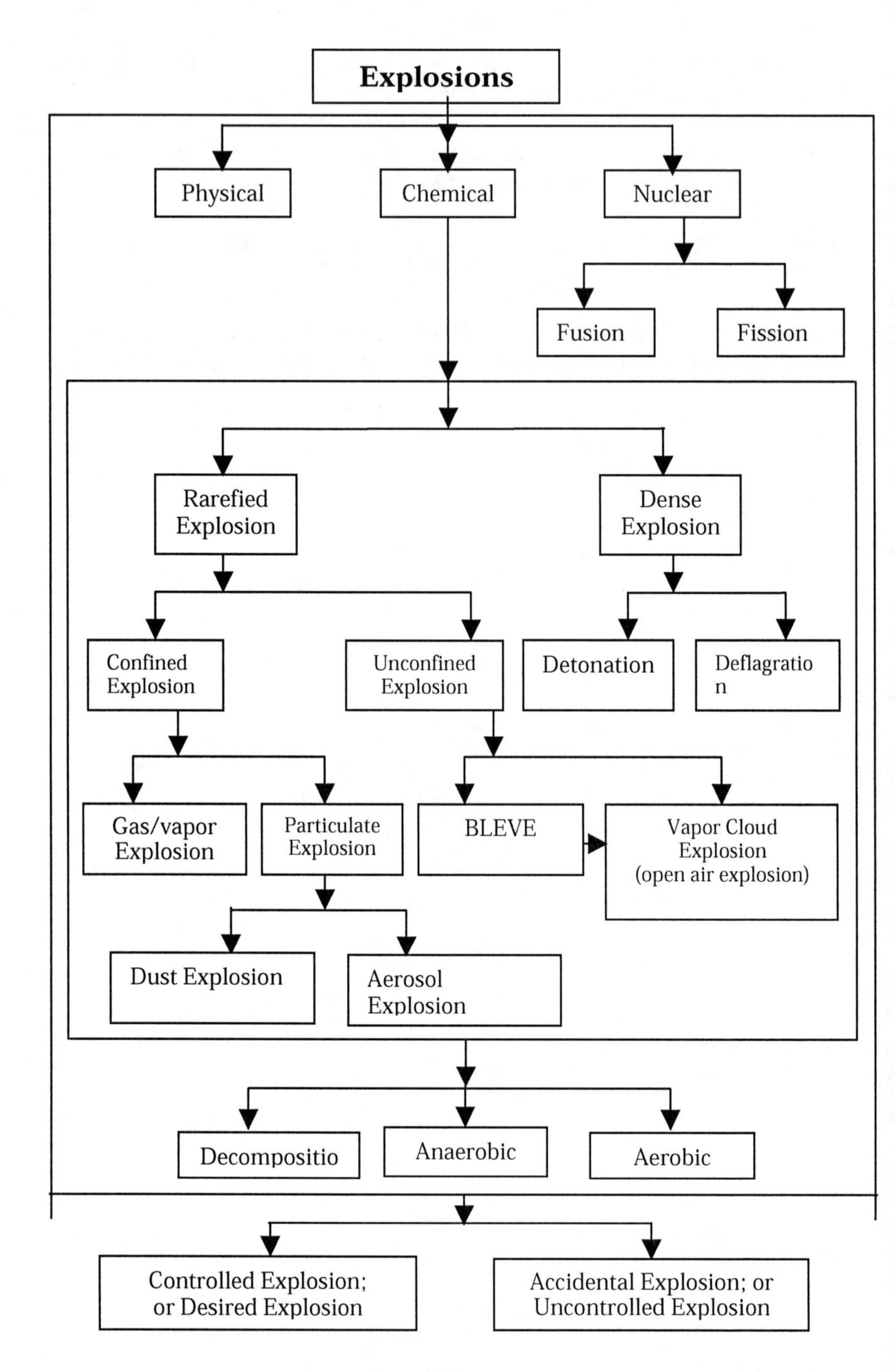
Explosions
Physical
Chemical
Nuclear
Fusion
Fission
Rarefied Explosion
Dense Explosion
Confined Explosion
Unconfined Explosion
Detonation
Deflagratio n
Gas/vapor Explosion
Particulate Explosion
BLEVE
Vapor Cloud Explosion (open air explosion)
Dust Explosion
Aerosol Explosion
Decompositio
Anaerobic
Aerobic
Controlled Explosion; or Desired Explosion
Accidental Explosion; or Uncontrolled Explosion

Faraday Case – A Lightning Protection System (LPS) where the area to be protected is enclosed by a heavy metal screen (like a birdcage) or continuous metallic structure with no unbonded metallic penetrations. On such a system, the lightning current flows on the exterior of the structure, not through the interior.

Synonym: Faraday-like Shield.

Faraday-like Shield – Synonym of Faraday Case.

Field Magazine – See *Magazine, Field.*

Fire – 1. The act of initiating an explosion. 2. The act of firing a weapon or artillery. 3. Chemical action producing heat, light, and flame. It is a process of chemical combination with oxygen of carbon and other elements constituting the substance being burnt.

Also see: C*ombustion.*

-fire, A – *Afire*: Blazing, Flaming, or on fire.

Fire, Accuracy of – See *Accuracy of Fire.*

Fire Alarm – An apparatus for giving a signal on the breaking out of a fire; or a signal given on the breaking out of a fire.

-Fire, All – See *All-Fire.*

Fire Apparatus – An apparatus designed and used for fighting or extinguishing fire, such as automobile fire engines, ladder trucks.

Fire Area – Fire area means one of various sections of a building that are separated from each other by fire-resistant walls.

Firearm – A weapon from which a shot is generally discharged by smokeless gunpowder. The term 'firearm' is usually applied to small arms.

Firearm, Small – Small Firearm: It means a firearm capable of being fired while held with hands.

Fire Arrow – An arrow bearing a flaming substance to set its mark afire.

Fire, Automatic – See *Automatic Fire.*

Fire, Back – See *Back fire.*

Fireball – 1. A fire that is burning sufficiently rapidly for the burning mass to rise into the air as a cloud or ball. 2. A ball of fire or something resembling such a ball, such as:

(a) a brilliant meteor that may trail bright sparks;
(b) a grenade or bomb fired properly; and
(c) the highly luminous cloud of vapor and dust created by a nuclear explosion

Fire Balloon – The balloon which is sent up, usually at night, with fireworks that ignite at a regulated height; or the balloon which is raised by the buoyancy of air heated by a fire placed in the lower part of the balloon.

Fireband – Fireband denotes a projected burning or hot fragment whose thermal energy is transferred to a receptor.

Fire Bar – Fire bar means a bar of boiler furnace or a grate.

Firebed – It is a layer of burning fuel as that in the furnace under a boiler.

Fire, Bengal – See *Bengal fire.*

Fire Blanket – A blanket of fireproof or flameproof material designed and used in smothering small fires.

-Fire, Blank Powder – See *Blank-fire Powder.*

Fire Block – Pieces of wood nailed horizontally between studs or joists to prevent the spread of fire and hot gases.

Fire Boat – Fireboat means a boat designed and equipped with pumps and other apparatus for fighting fire on or from the water.

Firebolt – Synonym of thunderbolt and lightning.

Fire Bomb – Synonym of *Incendiary Bomb.*

Fire Boss – The person who is responsible to examines a coal mine to determine whether firedamp is present, to search for fires caused by blasting and to check on the general safety of the mine.

Synonym: Gasman.

Fire Bottle, Chemical – See *Chemical Fire Bottle.*

Firebox – A chamber, such as that of a furnace or steam boiler, which contains a fire.

Firebrand – A projected burning or hot fragment whose thermal energy is transferred to a receptor.

Firebreak – A barrier of man-made or naturally occurring cleared land that stops the advancement of a forest fire or prairie fire.

Firebrick – Refractory bricks capable of sustaining high temperature without fusion, used for lining fireplaces, furnaces, etc.

Fire Bridge – The low separating wall generally of firebrick between the hearth and the grate in a reverberatory furnace.

Fire Brigade – A body of firefighters.

Firebug – The person whose duty is to patrol a metal mine looking for fire hazards and other dangers.

Fire, Categorization of – Categorization of Fires: Fires may be categorized into four groups as follows:

(1) Pool fire – a tank fire or a fire from a pool of fuel spread over the ground or water.
(2) Jet fire – a fire from the ignition of a jet of flammable material.
(3) Fireball and BLEVE (Boiling Liquid Expanding Vapor Explosion) – a fire resulting from the overheating of a pressurized vessel by a primary fire. This overheating raises

the internal pressure and weakens the vessel shell, until it bursts open and releases its contents as a large and very intense fireball.

(4) Flash Fire – a fire involving the delayed ignition of a dispersed vapor cloud which does not cause blast damage. The flame speed is not as high as in an unconfined vapor cloud explosion, but the fire spreads quickly throughout the flammable zone of the cloud.

Fire Check – A fine shallow crack caused by sudden heating in an unglazed ceramic body or a glass article.

Fire Chief – The head of a fire department.

Synonym: Fire Marshal.

Fire Clay – Clay that can withstand high temperatures without deforming. It approaches kaolin in composition, the better grades containing at least 35 percent alumina when fired. Fire clay is intended and used for crucibles, firebrick, and many refractory shapes.

Fire Cock – The cock used to furnish water for extinguishing fires.

Fire Control – Fire protection or extinction.

Firecracker – Synonym of *Cracker.*

Fire-cure – Curing something such as tobacco over open fires in direct contact with the smoke.

Fire Cut – The slanted cut in the end of a wood beam or joist resting in a masonry wall, which in case of a fire allows the wood to fall out without wrecking the wall.

Firedamp – 1. Any flammable gas or any flammable mixture of gases occurring naturally in a mine. The flammable gas or any flammable mixture of gases occur naturally in mines by decomposition of coal or other carbonaceous matter and which consists chiefly of methane.

2. The *explosive mixture* which is formed by the flammable gas, mainly methane, with air in a coal mine.

Fire Detonator – Synonym of *Plain Detonator.*

Fire Direction – Tactical employment of firepower including the selection of targets and the massing of fires.

Fire Direction Center – An element of an artillery command post which is comprised of a gunnery, communication personnel, and equipment, with the aid of which the commander exercises fire control and fire direction.

Fire Division Wall – The wall that subdivides a fire-resistive building to restrict the spread of fire.

Fire Door – 1. An automatic door designed and used to release the door under the influence of heat and permit its closing by gravity or by weights or other contrivances. Such a door is secured in the open position by a fusible link or thermostatically operated device.

2. The door or opening through which fuel is supplied to a furnace or stove.

Fire Drill – A practice drill with fire-extinguishing apparatus or in the conduct and manner of exit to be followed in case of fire.

Fire, EC Blank – Synonym of *EC Smokeless Powder.*

Fire, Effect of – Effect of Fire: A fire affects its surroundings primarily through the radiated heat that is emitted. If the level of heat radiation is sufficiently high, other flammable objects can be ignited. In addition, living organisms may be burned by heat radiation. The damage caused by heat radiation can be calculated from the dose of radiation received; a measure of the received dose is the energy per unit area of the surface exposed to the radiation over the duration of the exposure. Alternatively, the likely effect of radiation may be estimated by using the power per unit area received.

The radiation effect of a fire is normally limited to the area close to the source of the release (say within 200 ft/60 m). In many cases, this means that neighboring communities are not affected. However, there are some types of fire that could have a more extensive effect, such as a fire in a gas well or in a petroleum installation.

Fire Engine – An automotive truck equipped with a motor-driven pump, hose and fire-extinguishing agents.

Fire Escape – A device intended to facilitate escape from a burning building. It is generally a stairway made of steel attached to the outside of the building.

Fire-exit Bolt – A locking device fitted to the exit doors of public buildings designed to be released by pressure applied from the inside of the building.

Fire Extinguisher – It is a device designed and used to put out fires by cutting off the supply of oxygen in air necessary for combustion. In practice the term 'fire extinguisher' is used to mean a portable or wheeled apparatus for putting out small fires by ejecting fire extinguishing agents which may consist of water alone, water and chemicals such as soda-acid solutions or foam, or chemicals alone such as carbon tetra-chloride, carbon dioxide, or dry chemicals.

Also see: Fire Triangle.

Firefang – The term 'firefang' means to become excessively dry, overheated, and damaged consequent upon slow oxidative decomposition of organic matter.

Fire Fight – The term 'fire fight' is used to mean an exchange of fire between opposing military units as distinct from the fighting when the two forces close with each other such as during an assault.

Fire Fighting – The effort to extinguish or to check the spread of a fire.

Fire Finder – A device consisting of a map and a sighting instrument for determining, as from a fire tower, the location of a forest fire.

Fire, First – See *First Fire.*

Fire, Flash – See *Flash Fire.*

Fire-float – A boat used to aid ships afire.

Fire Form – An act of reshaping a rifle cartridge case by loading and firing it until it conforms to the chamber of the rifle for which the case is being prepared.

Fire, Greek – See *Greek Fire.*

Fire Grenade – See *Grenade, Fire.*

Fire Ground – An area in which *fire fighting* operations are carried on.

Fire Guard – 1. The person responsible to watch for the outbreak of fire, as in a forest region; or to extinguish small fires. 2. Synonym of Fire Screen and Fire Break.

Fire, Hang – See *Hang Fire.*

Fire, Indian – Indian Fire: Synonym of *Bengal Fire.*

Fire, Jet – Jet Fire: A fire from the ignition of a jet of flammable material.

Firelock – A g*unlock* employing a slow match to ignite the powder charge. Also a gun having such a lock.

Fire Lookout – A lookout stationed in a fire tower that keeps watch over a large area of forest, and on sighting a fire notifies a dispatcher of its location.

Synonym: Towerman.

Fire Main – A pipe for water to be used in putting out fire.

Fire Maker – A device once used to make fire. It consisted of a piece of flint held in place by metal prongs, and struck by a hammer like that of a musket by cocking that hammer and pulling a trigger. The operation used to produce spark and fell into a metal box filled with wood shavings or other flammable material causing it to ignite.

Fireman – 1. The crewmember of a steam operated locomotive whose principal duties include firing and operating the boiler. 2. In a locomotive which runs other than on steam, the crewmember whose responsibility is to assist the locomotive engineer by watching for signals or track obstructions. 3. The person whose profession is to fight fire.

Fire Marshal – The head of fire-prevention organization, or the person in charge of the fire-fighting personnel and equipment of a large establishment.

Synonym: Fire Chief.

-fire, Mis – See *Misfire.*

Fire Mission – Fire mission implies the assignment of a specific target and includes an order as to when to fire the amount of ammunition to be used.

-fire, No, Current – See *No-fire Current.*

Fire Partition – A fire-resistant interior wall intended to provide protection to occupants during the evacuation of a burning building, to retard the spread of fire.

Fire Patrolman – One whose duty is to patrol a specific area such as a mine, factory, or national forest, watching for fires or fire hazards.

Fire Pit – A pit, the floor of which is partly or wholly of incandescent lava.

Fireplug – Fireplug means a hydrant for drawing water from the mains in a street or building for the purpose of extinguishing fires.

Fire, Plunging – See *Plunging Fire.*

Fire Point – The fire point of a liquid is the lowest temperature at which the vapor of the liquid ignites and burns steadily in air. Generally, the fire point of a liquid is above its flash point.

Compare: *Ignition Temperature.*

Fire, Pool – Pool Fire: A tank fire or a fire from a pool of fuel spread over the ground or water.

Firepot – A vessel that holds fire.

Fire Powder, Blank – Blank Fire Powder: Synonym of *EC Smokeless Powder.*

Firepower – The term 'firepower' means the capability, as of a military unit, to deliver effectively and promptly missiles, shells, etc upon a specific target.

Fire Prevention – The measures taken and the practices done to prevent the outbreak of fire and to suppress destructive fires at a particular place.

Fireproof – Proof against fire. In the case of a building having all parts which resist stresses or carry weights, and also all interior and exterior walls and stairways made of noncombustible materials, the building is referred to as fireproof.

Fire Protection – Measures and practices for preventing or reducing injury and loss of life or property by fire. Fire protection measures include design features, systems or equipment.

Firer – One who fires, or one who lights, replenishes, and attends to a fire as for a brick kiln.

Also see: *Shot Firer.*

Fire Raft – A raft loaded with combustible materials intended for setting fire to an enemy's ships or waterfront.

Fire Resistance – The expression 'fire resistance' means the power of an element to withstand heating due to a fire of specified severity, while performing its normal functions, restricting the spread of fire in a building or to nearby structures for a definite time. Fire resistance denotes the length of time during which a construction offers resistance to a fire.

Fire Resisting Level – Synonym of *Fire Resisting Rating.*

Fire Resisting Rating – Usually referred to by its acronym FRR, it is a measure of the fire resistance of a material or structure. The minimum duration for which all sides of an element of structure, any of which is subjected to a standard fire, continues to perform its structural function and does not permit the spread of fire. Where a period of time is used in conjunction with the acronym FRR it is required that the element of structure referred to shall have a fire resistance rating of not less than the period stated.

A wall constructed of brickwork, concrete or masonry, 200 mm thick or reinforced concrete 150 mm thick, is deemed to have a four-hour fire resistance rating, referred to as FRR 4 hour.

Fire-Resistive – Synonym of *Fire-resisting.*

Fire-Resistive Door – Synonym of *Fire Door.*

Fire-Retardant – A substance having the ability or tendency to halt or slow the spread of fire by providing insulation or structural adequacy.

Fire Retarder – An element of structure protected with a fire-retardant material.

Firer, Shot – See *Shot Firer.*

Fire Runner – A person whose duty is to go into a mine after the blasting to search for fires and to replace damaged brattices.

Firesafe – A substance said to be firesafe that is offering protection against fire or is protected against fire.

Firesafety – A state of protection against fire.

Fire Screen – A protecting wire screen or grating placed ahead of or fitting over the front of an open fireplace.

Fire Service – 1. An organized fire-fighting and fire-preventing service of a country, town, etc. 2. The occupation of fire fighting.

Fire Ship – A ship carrying explosives or inflammables sent among the enemy's ships or works to set them on fire.

Fire Shutter – A metal shutter constructed to resist fire for a period often specified as one hour.

Fire Station – A building that accommodates fire apparatus and firemen.

Firestone – Pyrite formerly used for striking fire.

Synonym: Flint

Fire Stop – A member or material used to fill or close open parts of a structure for preventing the spread of fire and smoke.

Fire Storm – A tremendous inrush of air that may lead to hurricane force resulting from an extremely large area fire or a nuclear bombing.

Fire, Striking – See *Striking Fire.*

Fire Superiority – Higher capacity with respect to arms, ammunition and explosives in comparison to that of the enemy.

Also see: *Firepower.*

Fire Support – The assistance given by artillery fire, naval gunfire, and airplane strafing and bombing to infantry and armored units.

Fire Tower – A fireproof and smoke proof compartment attached to or running vertically through a building and having a fireproof stairway.

Fire Triangle – Fire is a chemical reaction between a fuel and oxygen in presence of heat. These three components that are essential for a fire to occur can be represented by the three arms of a triangle.

Fire triangle: Fuel + Oxygen + Heat Source = Combustion.

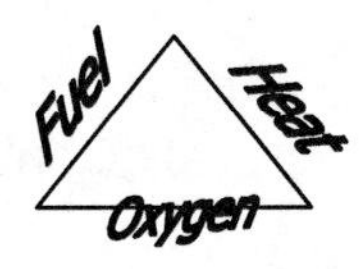

A fire cannot take place in absence of any of these three factors.

Fire triangle is a concept from the art/science of firefighting. In extinguishing a fire, all that is done is to break any of the arms of the fire triangle. In order to extinguish fire, heat may be removed and brought below the ignition point of the substance involved by cooling, the supply of oxygen may be stopped by smothering the fire in foam or dry sand (smothering), or the fuel may be removed from the scene of the fire (by isolation). Whatever may be the equipment or extinguishing media for fire fighting, cooling, smothering and isolation or starvation are the three processes for fire extinction.

Compare: *Explosion Triangle.*

Fire Wagon – Synonym of *Fire Engine.*

Firewall – The wall or other barrier designed, constructed and placed with the objective of preventing the spread of fire or the radiation of heat from any one place to some other place.

Fire Wind – A wind caused by a *firestorm.*

Fireworks – Fireworks are a specialized class of *pyrotechnic* and explosive devices designed and prepared to create audible or visible effects for entertainment. These are explosive devices that burn with colored flames. Fireworks may be a combustible or explosive composition or manufactured articles. Fireworks include firecrackers, skyrockets, smoke pots, whistlers, shells, and Roman candles.

Fireworks are usually composed of black powder plus various additive agents. For instance, the salts of strontium may be added to impart a red color to the flash, which occurs when the black powder is ignited. Powdered aluminum or other metals may be added to produce a greater amount of flame and sparks. At any rate these are deflagrating mixtures and so properly belong in the class of low explosives.

Fireworks are used not only for entertainment, but also by military, navy, etc. for illumination, marking and signaling.

Fireworks are assigned to one of following four types: *Fireworks, type A*; *Fireworks, type B*, *Fireworks, type C* and *Fireworks, type D.*

Also see: *Pyrotechnics.*

Fireworks, Low Level – See *Low Level Fireworks.*

Fireworks, Type A – Fireworks of such a nature that they have a risk of mass explosion when packed for transport.

Fireworks, Type B – Fireworks of such a nature that they do not have a risk of mass explosion when packed for transport but have a risk of projection of missiles. Such missiles include fragments of the casing of fireworks, pyrotechnic objects such as ejected stars and also self-propelled missiles such as rockets.

Fireworks, Type C – Fireworks that when packed possess a risk of fire and insignificant explosion hazard with no risk of projection. This type also includes small fireworks that would be Type B if the projection hazards were not largely prevented by the packaging.

Fireworks, Type D – Fireworks having no significant explosion risk when packed.

Firing Current – The electric current of recommended magnitude and duration used for the purpose of initiation of an electric detonator or a circuit of electric detonators.

Firing Current, Minimum Recommended – See *Minimum Recommended Firing Current.*

Firing Device – Firing device means any item designed and used to initiate by mechanical means a detonator or an igniter.

Firing Line – Synonym of Leading Wire. See *Wire, Leading.*

Synonym: *Leading Line*, *Lead Line*, *Leading Wire*, and *Lead Wire*.

-firing Operation Shot – Shot-firing Operation: See *Operation, Shot-firing.*

Firing Order – The order in which the several cylinders of an internal-combustion engine are sparked and fired.

Firing Pad – A specially prepared site wherein explosive items are fired for the purpose of test data acquisition.

Firing Pin – The needle-like metal part of a modern firearm that gives a vigorous strike to the primer, which initiates the firing of the cartridge.

Firing Point – A position on a firing line from which weapons are fired.

Firing, Premature – See *Premature Firing.*

Firing Ring – A member of a set of flat clay rings of different sizes used in a pottery kiln. The set of rings are so placed that they may be withdrawn successively as the firing proceeds. The amount of shrinkage of the ring indicates the intensity of the fire.

Firing, Rotational – See *Rotational Firing.*

Firing Site – An area for conducting test firing of explosives where access is restricted.

Firing Train – Synonym of Explosive Train. *See Train, Explosive.*

Synonyms: Explosive Train, and Initiation Sequence.

Also see: *Detonation.*

First Fire – A pyrotechnic term denoting the igniter composition used with pyrotechnic devices, which is loaded in direct contact with the main pyrotechnic charge. A first fire composition is compounded to produce a high temperature and hot slag. It is essentially readily ignitable and capable of igniting the underlying pyrotechnic charge.

Fission Bomb – See *Atom Bomb.*

Also see: *A-bomb, Atomic Bomb, Fusion Bomb,* and *Nuclear Weapon.*

Fixed Ammunition – See *Ammunition, Fixed.*

Fixed Round – See *Round, Fixed.*

Fizz Zone – See *Zone, Fizz.*

Flammability – The ease with which a material can be ignited by flame and heat.

Flammable – Capable of being ignited and burning in air.

Flammable Limit – The range of concentration of a flammable material in air above and below which combustion will not propagate. The limits between the minimum and maximum concentrations of vapor in air, which form flammable or explosive mixture, are usually abbreviated as LFL (Lower Flammable Limit) and UFL (Upper flammable Limit).

Also see: *Lower Flammable Limit,* and *Upper Flammable Limit.*

Flammable Liquid – Flammable Liquids are liquids, mixtures of liquids, or liquids containing solids in solution or suspension, which give off an inflammable vapor at temperatures of not more than 60.5° Celsius, closed cup test, or not more than 65.6° Celsius, open-cup.

Flammable Range – See *Flammable Limit.*

Flame – Flame is a glowing mass of gas produced during combustion. Flame is the state of blazing combustion. It is the output of a chemical reaction or is the reaction product, mostly gaseous, which yields heat and light. The reactants flow into and products flow out of the hot, luminous *flame front.*

Flame Arrestor – Any device or assembly designed and used to prevent the passage of flames into enclosed spaces. The device or assembly can be of cellular, tubular, pressure or other type.

Flame Front – The expression 'flame front' means the boundary between the burning and unburned portions of a pneumatic mixture of a flammable vapor and air, or other combusting system.

Flame, Jet – See *Jet Flame.*

Flame, Oxyhydrogen – Oxyhydrogen flame: The flame generated from the combustion of a mixture of oxygen and hydrogen.

Flameproof – 1. Literally, flameproof means incapable to sustain combustion. It means anything having resistance to the action of flame. 2. In relation to equipment, the term 'flameproof' means the enclosure of the equipment which is capable of withstanding an internal explosion, and prevents passage of flame to the surrounding atmosphere.

Flame Screen – It is a portable or fitted device incorporating one or more corrosion-resistant wire woven fabrics of a very small mesh used for preventing sparks from entering a tank opening or for a short period of time preventing the passage of a flame, yet permitting the passage of gas. Fitted single screen's mesh is essentially at least 11 per linear centimeter (28 per linear inch) and for two fitted screens the mesh must be at least 8 per linear centimeter (20 per linear inch), spaced not less than 1.25 centimeter (½ inch) or more than 3.7 centimeters (1-1/2 inches) apart.

Flame Speed – The rate at which combustion moves through an explosive mixture.

Flame Zone – See *Zone, Flame.*

Flare – An unsteady glaring light, or to burn with an unsteady or wavering flame. In effect it is a pyrotechnical device designed to produce a single source of intense light or radiation for relatively long durations for target or airfield illumination, signaling or other purposes. However, oil refinery or gas field flare is not a pyrotechnical device.

Flare, Aerial – See *Aerial Flare.*

Flare, Aluminum – See *Aluminum Flare.*

Flareless Explosion – Flareless Explosion means an *explosion* without a flame.

Flare, Magnesium – See *Magnesium Flare.*

Flare, Surface – Surface flare is a pyrotechnic device used to illuminate, identify, signal or warn.

Flash – 1. A sudden burst of flame or light. 2. The total lightning event. A flash may involve several lightning strokes, generally using the same path through the air as the initial event.

Flash Cartridge – It consists of a paper cartridge shell, a small arms primer and a flash composition, all assembled in one piece ready for firing.

Flash Fire – Flash fire means the combustion of a flammable gas or vapor and air mixture in which flame passes through that mixture at less than sonic velocity, such that negligible damaging overpressure is generated.

Flashless Explosive – See *Cool Explosive*, and *Nitroguanidine.*

Flash Over – The sympathetic detonation between explosive charges or between charged blast holes.

Flash Past – The possibility of an igniting flash bypassing a delay element in assemblies such as delay detonators and military fuses.

Flash Point – The flash point is the lowest temperature at which an inflammable liquid gives off sufficient vapor so as to form a momentary flash with air on application of a small flame in the prescribed manner in the flash point apparatus.

Compare: *Fire Point.*

Flash Powder – Flash powder is a mixture consisting of powdered aluminum or a magnesium/aluminum alloy, which upon ignition can result in a violent explosion and flash.

Flexible Detonating Cord – It consists of a core of detonating explosive enclosed in spun fabric with or without plastics or other covering and wire countering.

Flexible Linear Shaped Charge – See *Shaped Charge, Flexible Linear.*

Flex-x – A secondary high explosive. It is flexible, waterproof, and insensitive to shock. Mainly used in cutting charges. Rate of detonation is 6800 meters (22,300 feet) per second.

Flowers of sulphur – Sulphur in the form of powder.

Foam Zone – The initial stage of partial gasification in the burning of a propellant.

Forbidden Explosive – 1. In accordance with the regulations of the U.S. Department of Transportation (DOT), an explosive which is not acceptable for transportation by common, contract, or private carriers, by rail freight, rail express, highway, air or water.

2. The following are usually the forbidden explosives, namely:

(i) Ammunition loaded firearms;
(ii) An explosive composition that ignites spontaneously or undergoes marked decomposition when subjected to a temperature of 75°C for 48 hours;
(iii) An explosives containing a chlorate and also (a) an ammonium salt, including a substituted ammonium or quaternary ammonium salt, (b) an acidic substance, including a salt of a weak base and a strong acid;
(iv) Firecrackers and other fireworks which combine an explosive and a detonator;
(v) Fireworks that contain yellow or white phosphorus;
(vi) Leaking or damaged packages of explosives;
(vii) New explosives and explosive devices yet to be authorized;
(viii) Nitroglycerin, diethylene glycol dinitrate or any other liquid explosives;
(ix) Propellants that are unstable, condemned or deteriorated.

Formulation – 1. An explosives composition. 2. The operation of combining ingredients to produce a mixture of a final desired composition possessing specific physical and explosive properties.

Free Face – In blasting, a surface of coal or rock exposed to air or water, which provides room for expansion upon fragmentation. Usually the surface is approximately parallel to the line of boreholes.

Synonym: Open Face.

Fracturing Device for Oil Well – A device designed and used to fracture the rock around drill shafts in order to assist the flow of oil from the rock. The device consists of a metallic case containing a charge of detonating explosive with a detonator and an explosive chain, which is generally actuated by clockwork.

Fragmentation – The breaking and scattering the pieces of a bomb, grenade or projectile in all directions; or the breaking of a solid mass into pieces by blasting.

Friction Fuse – The fuse that is ignited by the heat evolved by friction.

FRR – Acronym for *Fire Resisting Rating.*

Fuel – Generally a substance that may react with oxygen to produce combustion. In an explosive, it is the ingredient that reacts with an oxidizer to form gaseous products of detonation. Any combustible such as sulfur or aluminum powder is a fuel in pyrotechnics.

Antonym: Oxidizer.

Fuel-Air Explosive – See *Vacuum Bomb.*

Fuel Oil – The fuel, usually diesel, in *ANFO.*

Fulminate – Fulminate means to explode with a vivid thunderstorm. It is in effect a salt or ester of fulminic acid, especially mercury fulminate, $Hg(ONC)_2$. Other than mercury fulminate, examples of fulminates commonly used in the explosives industry are copper fulminate and silver fulminate.

The fulminates are a class of explosives. These primary high explosives are very sensitive to flame, friction, and impact when dry.

They decompose completely when they detonate and do so with great violence. If compressed beyond 1700 kg/cm^2 (25,000 psi) they become what is known as "dead-pressed," or not capable of being exploded by flame. Fulminates are subjected to deterioration when stored in hot atmosphere.

They are widely used as initiators or primers for bringing about the detonation of high explosives or the ignition of powder. Fulminates are used for the caps or exploders by means of which charges of gunpowder, dynamite, and other explosives are fired. They are commonly used in combination with substances that provide a more prolonged blow and a bigger flame than fulminates alone. Such a material might be potassium chlorate. In the reinforced type of detonator, fulminates are made more effective by the addition of a more sensitive and powerful high explosive. This material was generally used in the manufacture of caps and detonators for initiating explosions for military, industrial, and sporting purposes, but has been banned. Primer caps and detonators loaded with fulminate of mercury are less sensitive than the dry bulk material but should still be handled with great care. Stocks in an assembly or loading room should be kept as small as possible.

Also see: *Fulminate of Mercury, Fulminating Mercury, Mercury Fulminate, Fulminate 1.*

Fulminate, Copper – See *Copper Fulminate.*

Fulminate, Mercury – See *Fulminate of Mercury.*

Fulminate, Mercuric – See *Fulminate of Mercury.*

Fulminate of Gold. – See *Gold, Fulminate of.*

Synonym: Fulminating Gold.

Fulminate of Mercury – Chemical name is mercuric isocyanate, $Hg(ONC)_2$. The *fulminate* when dry explodes violently if struck or heated. It is used in detonators and blasting caps and percussion caps.

Synonyms: *Fulminating Mercury, Mercury Fulminate.*

Also see: Blasting Cap, Fulminate, and Percussion Cap.

Fulminating Compound – An early term, applied to mixtures of chemicals generally containing silver and used essentially for pyrotechnic purposes.

Fulminating Gold – See *Gold, Fulminating.*

Synonym: Fulminate of Gold.

Fulminating Mercury – See *Fulminate of Mercury.*

Fulminating Powder – Synonym of *Percussion Powder.*

Fulminating Silver – A black substance which when dry is an explosive. It may be a nitride of silver (Ag_3N). It is formed when a solution of silver oxide in ammonia is exposed to the air.

Fume – The gaseous products of an explosion especially in mining. In practice, poisonous or toxic gases such as carbon monoxide, hydrogen sulfide and nitrogen oxides are considered as fumes.

Fuse – 1. An igniting or explosive device in the form of a cord or string designed and used to initiate a continuous train of explosives by deflagration or detonation. It consists of a flexible fabric tube and a core of low or high explosive. Used in blasting and demolition work, and in certain munitions. In short, a fuse is a cord that contains a core of black powder that burns at a specific rate and is used to initiate a blasting cap.

A fuse with a black powder or other low explosive core, which burns at a specific rate, is called a '*safety fuse*' or '*blasting fuse*'. A fuse with a PETN or other high explosive core is called '*detonating cord*' or '*prima cord*'.

The blasting or safety fuse, employed to fire an explosive from a distance or after a delay, is designed to propagate burning at a slow and steady rate. The far end of the fuse is usually embedded in the explosive charge. Detonating or prima cord is fired by a detonator and is capable of initiating the detonation of certain other explosives at any number of points and in any desired pattern.

This device transmits the flame to the detonator or charge either after an interval of time, or by an operation conducted at a distance. In either case the shot firer is not exposed to the effects of the explosion.

2. A safety mechanism used to prevent excessive current in a circuit and consequent overheating of wires. An easily fusible conductor inserted in series, and melting at definite amperage.

Antonyms: Defuse.

Synonyms: Fusee, Primer, and Priming.

Also see: Fuse.

Fuse, Bickford – See *Bickford Fuse.*

Fuse, Blasting – S*ee Blasting Fuse*.

Fuse Cap – A detonator that is initiated by a safety fuse.

Synonym: Fuse Detonator, Ordinary Blasting Cap.

Fuse, Capped – See *Capped Fuse.*

Fuse, Cordeau Detonant – Cordeau Detonant Fuse: A term used to define detonating cord.

Fuse Cutter – A mechanical device used to cut safety fuse so as to keep it clean and at right angles to its long axis.

Fuse, Delay – Delay Fuse: Any fuse incorporating a means of delaying its action. Delay fuses are classified according to the length of time of the delay.

Fuse, Detonating Metal Clad – Metal Clad Detonating Fuse: See *Metal Clad Detonating Cord.*

Fuse Detonator – A detonator that is initiated by a safety fuse.

Synonym: *Fuse Cap,* and *Ordinary Blasting Cap.*

Fusee – Fusee is a type of flare or a special type of match. The term is also synonym of fuse.

Fuse, Electric – Electric Fuse: The term implies the firing of explosives by means of an electric current. Heat generated by a current is utilized to fire a suitably constructed fuse, and this ignites a detonator, which in turn explodes the charge. Electric fuse consists of a metallic cup containing primary explosives, in which two insulated conducting wires are held fixed by a plug, and the ends of the wires are held near to but not touching each other in the latter. A little amount of sensitive priming is kept in the plug. As and when an electric current is passed through these conductors a spark fires the priming followed by the primary explosives and the main charge explosive.

Fusehead – Fusehead is the ignition element in an electric detonator.

Fuse Hole – The expression 'fuse hole' means the hole in a shell prepared for the reception of the fuse.

Fuse Igniter – Fuse Igniter means small cylindrical hollow metal tubes or pasteboard containing an igniting composition in one end, the other end being open.

Fuse, Igniter, Tubular, Metal Clad – Tubular Metal Clad Igniter Fuse: It consists of a metal tube with a core of deflagrating explosive.

Fuse, Igniter – See *Igniter Fuse.*

Fuse, Instantaneous, Non-detonating – Cotton yarns impregnated with meal powder.

Fuse Lighter – It is a pyrotechnic device for the rapid and dependable lighting of safety fuse. A lighter is a small hollow pasteboard or a metal tube with either a wick or two wires connected to a small quantity of igniting composition. Safety lighters may or may not be electrically actuated.

Fuse Plug – In relation to ordnance, a plug fitted to the fuse hole of a shell to hold the fuse.

Fuse Powder – A very fine black powder of mesh 140 or more, which is used for safety fuses.

Fuse, Safety – See *Safety Fuse.*

Fuse,Time – See *Time Fuse.*

Fuse, Tracer – See *Fuse Tracer.*

Fusible Link – A safety device consisting of a suitable low melting point material, which is intended to yield or melt at a predetermined temperature.

Fuse, Tracer – See *Tracer Fuse.*

Fuse/Fuze* – Although the two words 'fuse' and 'fuze' have a common origin and are sometimes considered to be synonyms, it is useful to maintain the convention that "fuse" refers to an igniting or explosive device in the form of a cord or tube whereas "fuze" refers to a device which incorporates mechanical, electrical, chemical or hydrostatic components to initiate a train by deflagration or detonation. [*Quoted from UN code.]

Fusion Bomb – Synonym of *Hydrogen Bomb.*

Also see: *Nuclear Weapon.*

Fuse – A fuse is an essential and critical part of effective munitions. It is a device that incorporates mechanical, electrical, chemical or hydrostatic components to initiate a train by deflagration or detonation. It is generally a detonating device for initiation of, as by percussion, the bursting charge of bomb, projectile, or torpedo. A fuse is used to ensure safe, reliable detonation of munitions at the desired place and time. It is thus a defense-unique product with little use in non-military applications.

Synonyms: Fusee, Primer, and Priming.

Fuse, Bomb – See *Bomb Fuse.*

Fuse, Concussion – See *Concussion Fuse.*

Fuse, Confined Detonating – See *Confined Detonating Fuse (CDF).*

Fuse, Delay – Delay Fuse: A fuse incorporating a means of delaying its action. Delay fuses are classified according to the length of time of the delay.

Fuse, Delay, Short – See Short Delay Fuse.

Fusee – 1. A wooden match with a bulbous head not easily blown out when ignited. 2. A paper match impregnated with niter and tipped with sulfur. 3. A red signal flare used especially for protecting stalled trains and trucks.

Fuse, Electric – Electric Fuse – A fuse which is ignited by heat or a spark produced by an electric current.

Fuse, Friction – Friction Fuse: A fuse that is ignited by the heat evolved by friction.

Fuse, Impact – Impact Fuse: Fuse designed to function on impact.

Fuse, Instantaneous – Instantaneous Fuse: One that will burst the projectile on the outside of a hard surface (such as a concrete emplacement) before penetration or ricochet. This fuse creates some crater on hard ground.

Synonym: Super Quick Fuse.

Fuse, Super Quick – Super Quick Fuse: Synonym of *Instantaneous Fuse.*

Fuse, Detonating – A device with explosive components designed and used to produce a detonation in ammunition or commercial explosives.

Fuse, Point Detonating – See *Point Detonating Fuse.*

Fuse, Igniting – Igniting Fuse: A mechanical device with explosive components designed and used to produce a deflagration in ammunition.

Fuse, Impact – See *Impact Fuse.*

Fuse, Long Delay – Long Delay Fuse: A type of delay fuse in which the fuse action is delayed for a relatively long period of time. The delay time may be from minutes to days, depending upon the type.

Fuse, Medium Delay – See *Medium Delay Fuse.*

Fuse, Mild Detonating – See *Mild Detonating Fuse.*

Acronym: *MDF.*

Fuse, Non-Delay – See *Non-Delay Fuse.*

Fuse, Proximity – See Proximity Fuse.

Fuse, Short Delay – See *Short Delay Fuse.*

Fuse, Supersensitive – Supersensitive Fuse: A fuse that sets off a projectile when it strikes even a very light target, such as an airplane wing.

Fuse, Time – Time Fuse: It is a mechanism for igniting the bursting charge of a projectile at some predetermined time after the projectile is fired.

Gaine – A booster charge of intermediate range used between a detonator and an insensitive high explosive.

Gallery – Equipment used for firing an explosive into an incendive mixture of air and methane.

Galvanometer, Blasting – See *Blasting Galvanometer.*

Gap Sensitivity – It is a measure of the distance up to which an explosive can propagate the detonation. It is used for measuring the likelihood of sympathetic detonation.

Gas Calorimeter – Equipment designed and used to determine the calorific value of a fuel gas by burning at a known rate or, for small amounts of combustible gas, combustion or explosion of a known volume of gas.

Gas, Detonating – See *Detonating Gas.*

Gas Detonation System – A system for initiating cap in which the energy is transmitted through the circuit by means of gas detonation inside a hollow plastic tube.

Gas Explosion – The rarefied explosion that may occur from *explosive mixture* that is formed by a flammable gas with air.

Gasless Delay – Delay elements consisting of a pyrotechnic mixture that burns without production of gases.

Gasless Delay Detonator – The original name for modern delay detonator.

Gas Free – An atmosphere in the tank, compartment or receptacle that has been tested using an appropriate gas detector and found to be sufficiently free, at the time of the test, of flammable gases or vapors for a specified purpose. For purpose of man-entry, the concentration of toxic gas in the tank, compartment or receptacle must be below the threshold limit but the atmosphere has to have the minimum oxygen content unless breathing apparatus is provided.

Gas, Military – Military Gas: A military gas is any agent or combination of agents that is capable of producing either a toxic or irritating physiological effect. Such agents may be in solid, liquid, or gaseous state, either before or after dispersion. Military gases may be classified as persistent and non-persistent. If it remains effective at its point of release for more than 10 minutes, it is termed as persistent. If it becomes ineffective within 10 minutes, it is classed as non-persistent.

Gas, Mustard – Mustard Gas: A blister gas that acts as a cell irritant and cell poison. It contains about 30 percent sulfur impurities, giving it a pronounced odor.

Gassy Coal Mine – A coal mine in which methane may be present.

Synonym: Safety Lamp Mine.

Gel – Jelly formed from a colloidal suspension dispersed in a liquid.

Gelatin – An explosive or blasting agent having a gelatinous consistency. It is in effect an explosive that is a jelly of nitroglycerine containing nitrocellulose, normally with oxidizing salts and solid fuel dispersed in it. The term is generally applied to gelatin dynamite but may also be a water gel.

Gelatin, Ammonia – Ammonia Gelatin: A gelatinous explosive or blasting agent containing nitroglycerin, nitroglycol, or similar liquid sensitizers and nitrocellulose with a portion of the liquid sensitizer replaced by ammonium nitrate. It has a very good water resistance quality.

Gelatin, Blasting – See *Blasting Gelatin.*

Gelatin Dynamite – A type of highly water-resistant dynamite characterized by its gelatinous consistency. This dynamite is gelatinized with nitrocellulose to form a plastic cohesive mixture. Due to its waterproof and plastic properties it can be loaded solidly into boreholes. Does not release much poisonous gas on detonation, and is therefore suitable for underground work

Gelatin, Special – Special Gelatin: The *gelatin explosive* in which the main oxidizing ingredient is *ammonium nitrate.*

Gelatin, Straight – See *Straight Gelatin.*

Gelignite – An explosive comprised of a mixture of nitroglycerin, nitrocellulose, potassium nitrate and wood pulp.

It is in effect a type of dynamite in which the nitroglycerin is absorbed in a base of wood pulp and sodium or potassium nitrate.

Also see: *Dynamite.*

Gelignite, Ammon – See *Ammon Gelignite.*

Gelignite, NS – NS Gelignite: A nitroglycerine gelatin explosive containing sodium nitrate as its main oxidizing ingredient.

Gel, Independent – See *Independent Gel.*

Gelly – A type of *dynamite* in which the *nitroglycerin* is absorbed in a base of wood pulp and potassium or *sodium nitrate.*

Synonym: *Gelignite.*

Gel, Water – See *Water Gel.*

Ghost Vapor – Water-laden Solvent Vapor.

Giant Low-Flame Powder – A type of the earlier American permissible dynamites which usually consisted of nitroglycerine, magnesium sulfate $MgSO_4$, $7H_2O$ and wood meal or kieselguhr. The water of crystallization of the magnesium sulfate is vaporized on explosion and cools the gases.

Also see: *Hydrated Explosive.*

Giant Powder – Trade name for dynamites made in the USA.

Glonoin – Before the invention of dynamite by Alfred Nobel, Nitroglycerin used to be used as a headache remedy under the name glonoin.

Glyceroltrinitrate – Synonym, rather actual name of what is known as *Nitroglycerin.*

Glyceryl Nitrate – Glyceryl nitrate is the correct chemical name of what is traditionally called *nitroglycerine* for general purpose.

Glycol Dinitrate – Synonym of *Nitroglycol*, *Ethylene Dinitrate*, and *Ethylene Glycol Dinitrate.*

Gold, Fulminate of – Fulminate of Gold: The explosive compound of gold is formed when auric hydroxide is heated with ammonia solution. It is a green powder. When dry it detonates violently when struck.

Synonym: Fulminating gold.

Grain – A system of weight (actually mass) measurement for propellant powder and bullets. 15400 grains equal one kilogram (7000 grains in a pound). Typical weights of gunpowder used in one cartridge range from about 20 grains to 80 grains depending on the cartridge type, powder type, bullet weight, and desired muzzle velocity. Typical .30 caliber bullets range from 125 to 220 grains. .22 caliber bullets for center-fire cartridges range from about 50 grains to 70 grains.

Grain, Multiple – See *Multiple Grain.*

Goop – Mixture of inflammable hydrocarbons, such as gasoline and magnesium powder; used in incendiary bombs.

Granular Dynamite – See *Dynamite, Granular.*

Greek Fire – An explosive mixture that catches fire when wet, used by the ancient Greeks in naval warfare. A Byzantine fleet first used it against an Arab fleet in about 668 A.D. It was then probably composed of sulfur, naphtha, and quicklime or similar materials.

This is essentially a mixture of nitrates and pitch, which has been in use since the 8th century. The material, being a deflagrating mixture, belongs to low explosive class.

Grenade – A small explosive bomb thrown by hand or fired from a missile. It is constructed of a hollow ball or shell of iron fitted with a priming charge and a bursting charge, and filled with a destructive agent, such as gas, high explosive, incendiary chemicals.

Also see: *Hand Grenade, Fire Grenade, Operational Grenade, Practice Grenade,* and *Rifle Grenade*.

Grenades, Hand – Hand Grenade: A grenade designed to be thrown by hand.

Grenade, Fire – Fire Grenade: A type of portable fire extinguisher, which is thrown into the flames. It consists of a glass bottle containing water and gas.

Grenade, Molotov – Molotov Grenade: Synonym of *Molotov Cocktail.*

Grenade, Operational – Operational Grenade: A grenade that contains a bursting charge.

Grenade, Practice – Practice Grenade: A grenade that contains a priming device and may contain a sporting charge.

Grenade, Primed Empty – See *Primed Empty Grenade.*

Grenades, Rifle – See *Rifle Grenade*.

Grioutite – A type of the earlier French and Belgian permissible/permitted dynamites, which usually consisted of nitroglycerine 42%, magnesium sulfate $MgSO_4$, $7H_2O$ 46% and wood meal or kieselguhr 12%. The water of crystallization of the magnesium sulfate is vaporized on explosion and cools the gases. Similar American permissible dynamites were called giant low-flame powder, and hydrated explosives.

Ground Ring Electrode – Ground Ring Electrode (GRE) means an earth electrode system that encircles the structure, either on or buried in the earth.

Ground Rod – A component of one type of earth electrode system, generally a cylindrical device driven into the soil. The ground rod is attached to the down conductor and serves the purpose of dissipating the current of a lightning flash into the earth.

Ground Terminal – A component of lightning protection system. Ground rods and buried metal plates are typical examples of earth electrode system. It transfers the current of a lightning flash to the earth. It is in direct contact with the earth and is connected to the down conductor.

Synonym: *Earth Electrode System.*

Grounded – Connected to earth or to some conducting body that serves in place of the earth.

Grounding – The term 'grounding' means providing an electrical path to the earth, generally to the earth electrode system.

Synonym: Earthing.

Group, Explosophore – See *Explosophore Group.*

Guanidine Nitrate – A high explosive that is stable, flashless and non-hygroscopic. It is used as a blasting explosive in combination with charcoal and inorganic nitrates.

Guanylnitrosoamino-guaanyltetrazene – It is used as a high explosive that produces fumes. It is used in priming compositions and sometimes in combinations with lead azide to lower the flash point of the azide.

Synonym: *Tetracine.*

Guhr Dynamite – See *Dynamite, Guhr.*

Guillotine – Guillotine is an explosive device that is designed and used to cut by driving a hardened knife through a cable or line.

Gun – Gun is a portable *firearm* such as pistol, rifle, revolver, and shotgun. It is, in fact, a piece of ordnance used to throw a projectile by the force of propellant explosive, and which consists of a barrel or tube closed at one end where the projectile is placed in front of the propellant charge to be ignited.

Also see: *Small Firearm.*

Gun, BB – See *BB Gun.*

Guncotton – Guncotton is a form of *nitro-cellulose* containing more than 12.6% of nitrogen. The discovery of guncotton was the first great advance in the chemistry of explosives when in 1846 Professor Christian Friedrich Schönbein of the University of Basel invented it. Guncotton was the precursor to nitrocellulose, or modern *gunpowder. Smokeless* gunpowder and most of the violent propellants contain guncotton. The heat developed by the explosion of 1 gram of guncotton is 1,010 calories.

Dry guncotton is one of the most dangerous explosives, as when dry and warm it is very susceptible to explosion by friction.

Guncotton began to replace black powder as a propellant around the time of the U.S. Civil War in the 1860s.

Also see: *Cellulose Nitrate*, and *Nitro-cellulose.*

Guncotton, Military – See *Military Guncotton.*

Gun, Drogue – Drogue Gun: An explosive device intended to eject a slug that is attached to a drag parachute or similar device.

Gunlock – A gun having a lock that employs a slow match to ignite the powder charge.

Gunpowder – An explosive that burns rapidly to produce hot pressurized gases capable of propelling bullets from guns. Gunpowder is used as a propellant for firearms. Gunpowder, being a low explosive, burns producing a subsonic deflagration wave rather than a supersonic detonation wave, as do high explosives. *Black powder* and *smokeless powder* are the two basic types of gunpowder.

Smokeless powder is used in almost all modern guns. Black powder, which is composed of a mixture of saltpeter (potassium nitrate), sulfur, and charcoal, had been the only military explosive available as a propellant for 550 years. Now it has almost been replaced by *smokeless powder.*

Black Powder produces a cloud of white noxious smoke when it burns. Smokeless powder burns much cleaner but may still produce a small puff of smoke. Almost all modern firearms use smokeless powder, not only because there is less smoke, but also because the bullets can be made to exit the gun faster.

The story behind first naming the gunpowder is as follows:
British scientist Roger Bacon (1214-1294) became aware of the discovery of explosives by Asian alchemists more than one thousand years ago when they discovered that mixtures of saltpeter (KNO_3) and sulfur could be exploded. This prompted Roger Bacon to experiment with mixtures of saltpeter, sulfur, and a new ingredient – carbon, in the form of charcoal. Bacon had made black powder. One hundred years later Berthold Shwartz experimented with black powder. He took a long iron tube and closed one end except for a tiny hole. He filled the tube with black powder and stuffed a small pebble in it. He touched a flame to the tiny hole and the pebble shot through the air with great speed. Schwartz had thus invented the 'gun' and hence the name 'Gunpowder'.

Gunpowder, Smokeless – See *Smokeless Gunpowder.*

Gun, Submachine – Submachine Gun. An easily portable automatic weapon that can fire over 500 cartridges per minute.

Gun, Tommy – Tommy Gun: Synonym of *Submachine Gun.*

Hand Grenade – A grenade designed to be thrown by hand.

Hand Signal Device – Hand held devices that produce visual signals, including highway flares, small marine distress flares and railway fusees, containing pyrotechnic compositions and designed to signal or warn by means of flame or smoke.

Hangfire – The detonation of an explosive charge at a time after its designed firing time. For a few seconds hangfire cannot be distinguished from a complete failure or misfire. It happens due to temporary failure or delay in the action of a primer, igniter or propelling charge. Hangfire is a source of serious fire accidents unless dealt with caution.

Hazard – Hazard is such a physical situation that poses a potential for injury or death of men or other animals, damage to property, damage to the environment or some combination of these.

Hazard Classification, Explosive – Explosive Hazard Classification: The UNO system consists of nine classes of dangerous materials, with explosives designated as Class 1. Explosives are by definition, by international classification, by experience and by any measure of assessment, inherently hazardous.

The explosives *hazard* class is further subdivided into six divisions, which are used for segregating ammunition and explosives on the basis of similarity of characteristics, properties, and accident effects potential. The six hazard class/divisions are described below in short in the tabular form and in details. For explosives, the term 'hazard class' should be used when referring to hazard class and division. On the basis of the hazards explosives are classified as follows:

Table: Explosives hazard class/divisions.

Hazard class/division	Hazard description
1.1	Mass explosion
1.2	Non-mass explosion, fragment-producing
1.3	Mass fire, minor blast or fragment
1.4	Moderate fire, no blast or fragment
1.5	Explosive substance, very insensitive (with a mass explosion hazard)
1.6	Explosive article, extremely insensitive

Division 1.1 Explosives: Consists of explosives that have a mass explosion hazard. A mass explosion is one that affects almost the entire load instantaneously.

Division 1.2 Explosives: Consists of explosives that have a projection hazard but not a mass explosion hazard.

Division 1.3 Explosives: Consists of explosives that have a fire hazard and either a minor blast hazard, a minor projection hazard, or both but not a mass explosion hazard.

Division 1.4 Explosives: Consists of explosives that present a minor explosion hazard. The explosive effects are largely confined to the package and no projection of fragments of appreciable size is to be expected. An external fire must not cause virtually instantaneous explosion of most of the contents of the package.

Division 1.5 Blasting Agents: Consists of very insensitive explosives. This division is comprised of substances that have a mass explosion hazard but are so insensitive that there is very little probability of initiation or of transition from burning to detonation under normal conditions of transport.

Division 1.6 Explosives: Consists of extremely insensitive articles that do not have a mass explosive hazard. This division is comprised of articles which contain only extremely insensitive detonating substances and which demonstrate a negligible probability of accidental initiation or propagation.

Hazard, Explosion – See *Explosion Hazard.*

Hazard, Major – See *Major Hazard.*

Hazard, Major Accident – See *Major Accident Hazard.*

Hazardous Location – Hazardous locations are commonly defined as follows:

Class I, Zone 0: A location where ignitable concentrations of flammable gases, vapors or liquids are present continuously; or are present for long periods of time. Equipment intended for use in a Class I, Zone 0 area is usually of the intrinsically safe, 'ia' type.

Class I, Zone 1: A location where ignitable concentrations of flammable gases, vapors or liquids are likely to exist under normal operating conditions; may exist frequently because of repair or maintenance operations or leakage; or may exist because of equipment breakdown that simultaneously causes the equipment to become a source of ignition; or are adjacent to a Class I, Zone 0 location from which ignitable concentrations could be communicated. Equipment intended for use in a Class I, Zone 1 area is usually of the flameproof, purged/pressurized, oil immersed, increased safety, encapsulated or powder filled type.

Class I, Zone 2: A location where ignitable concentrations of flammable gases, vapors or liquids are not likely to occur in normal operation or, if they do occur, will exist only for a short period; where volatile flammable liquids, or flammable gases or vapors exist, but are normally confined within closed containers where ignitable concentrations of gases, vapors, or liquids are normally prevented by positive mechanical ventilation; adjacent to a Class I, Zone 1 location from which ignitable concentrations could be communicated. Equipment that is intended for use in a Class I, Zone 2 area is usually of the nonincendive, non-sparking, restricted breathing, hermetically sealed or sealed device type.

Class I, Division 1: A location where ignitable concentrations of flammable gases, vapors or liquids can exist under normal operating conditions; may exist frequently because of repair or maintenance operations or because of leakage; or may exist because of equipment breakdown that simultaneously causes the equipment to become a source of ignition Equipment intended for use in a Class I, Division 1 area is usually of the explosion proof or purged pressurized type.

Class I, Division 2: A location where volatile flammable liquids or flammable gases or vapors exist, but are normally confined within closed containers; where ignitable concentrations of gases, vapors or liquids are normally prevented by positive mechanical ventilation; or adjacent to a Class I, Division 1 location, where ignitable concentrations might be occasionally communicated. Equipment intended for use in a Class I, Division 2 area is usually of the nonincendive, non-sparking, purged/pressurized, hermetically sealed, or sealed device type.

Class II, Division 1: A location where ignitable concentrations of combustible dust can exist in the air under normal operating conditions; ignitable concentrations of combustible dust may exist because of equipment breakdown that simultaneously causes the equipment to become a source of ignition; or electrically conductive combustible dusts may be present in hazardous quantities. Equipment intended for use in a

Class II, Division 1 area is usually of the dust-ignition-proof, intrinsically safe, or pressurized type.

Class II, Division 2: A location where combustible dust is not normally in the air in ignitable concentrations; dust accumulations are normally insufficient to interfere with normal operation of electrical equipment; dust may be suspended in the air as the result of infrequent malfunctioning of equipment; or dust accumulation may be sufficient to interfere with safe dissipation of heat or may be ignitable by abnormal operation. Equipment intended for use in a Class II, Division 2 area is usually of the dust-tight, nonincendive, non-sparking, or pressurized types.

Class III, Division 1: A location where easily ignitable fibers or materials producing combustible flyings are handled, manufactured or used. Equipment intended for use in a Class III, Division 1 area is usually of the dust tight or intrinsically safe type.

Class III, Division 2: A location where easily ignitable fibers are stored or handled. Equipment intended for use in a Class III, Division 1 area is usually of the dust tight or intrinsically safe type

Hazardous Substance – It is a substance, which by virtue of its intrinsic chemical properties, constitutes a *hazard*. A substance may also pose a hazard under action of temperature and pressure, or some combination of the intrinsic chemical properties and the physical properties. For example, air or water alone do not constitute a hazard and hence neither would be classed as a hazardous substance, but it may pose a hazard if compressed and heated.

Hazardous Substances, Classification of – Classification of Hazardous Substances: Hazardous Substances are classified into nine classes:

Class 1 – Explosives;
Class 2 – Gases (compressed, liquefied or dissolved under pressure or deeply refrigerated);
Class 3 – Flammable Liquids;
Class 4 – Combustible Solids;
Class 5 – Oxidizing Substances;
Class 6 – Poisonous Substances;
Class 7 – Radioactive Substances;
Class 8 – Corrosives;
Class 9 – Miscellaneous hazardous Substances.

H-bomb – Synonym of *Hydrogen Bomb.*

H.E. – Abbreviation for *High Explosive.*

HEAT – Acronym for High Explosive Anti-Tank.

Heating Limits – The conditions, such as heating time, heating rate, maximum temperature, etc., established for safely heating an explosive system. Based on the estimated critical temperature of the explosive system, the heating limits are determined with a suitable margin of safety.

Heat of Explosion – It is the heat evolved in exploding (burning) a sample in a combustion bomb in an inert atmosphere under standard conditions of pressure and temperature.

Synonym: Explosion Heat.

Heat Test – See *Abel Heat Test.*

HEP Shell – Abbreviation for High-Explosive Plastic Shell.

Her/His Majesty's Explosive – See *HMX.*

Hess Test – A German test for brisance of explosives. The test is now little used.

Heterogeneous Explosion – An explosion that occurs from heterogeneous two-phase mixture. When very fine solid suspended particles of a combustible material are mixed with air in certain proportions, they form an explosive mixture that may explode when it comes in contact with a flame or spark. There are two phases in the explosive mixture, fuel is a solid and the oxidizer is gaseous material. It is a type of *rarified* chemical explosion.

Also see: *Aerosol Explosion*, and *Dust Explosion.*

Compare: *Homogeneous Explosion.*

Hexamine – Synonym of *Hexanitrodiphenylamine.*

Hexanitrodiphenylamine – A powerful and violent explosive used as a booster explosive. As a booster explosive it is superior to TNT. It is not as good a booster explosive as tetryl, but it is extremely stable and much safer to handle.

Synonyms: *Hexil, Hexite, Hexamine.*

Hexanitrodiphenylamino Ethyl Nitrate – This explosive is slightly less sensitive to impact than tetryl, but equally sensitive to detonation by other means. It is considered as powerful as tetryl. Its explosive strength can be enhanced by addition of oxygen-rich salts, such as potassium chlorate. It may replace tetryl as a booster as well as in detonating compositions.

Hexanitrodiphenylethylene – See *HNS.*

Hexanitrodiphenyloxide – This high explosive is stable and not very sensitive. It is used in detonating compositions and is considered more powerful than picric acid.

Hexanitrohexaazaisowurtzitane – A nitramine explosive invented in recent time. The high-energy explosive compound is about 20% more powerful than HMX.

Antonym: HNIW.

Synonym: CL20.

Hexanitromannitol – See *Mannitol Hexanitrate.*

Hexanitrooxanalite – Hexanitrooxanalite is about as powerful an explosive as TNT.

Hexanitrodiphenyl Sulfide – It is a powerful explosive and is used as a detonating explosive. This material has an added military advantage in that its explosion gases contain irritating and very toxic sulfur dioxide.

Synonym: Picryl sulfide.

Hexanitrodiphenyl Sulfone – A very powerful high explosive, used in detonating compositions. It is a stable material.

Hexatonal, Cast – Cast Hexatonal: *Hexogen* mixed with trinitrotoluene (TNT) and aluminum.

Hexil – Synonym of *Hexanitrodiphenylamine.*

Hexite – Synonym of *Hexanitrodiphenylamine.*

Hexogen – Synonym of *Cyclotricethylenetrinitramine or Cyclonite* or *R.D.X.*

Hexolite – An intimate mixture of Hexogen (Cyclotricethylenetrinitramine–RDX) and trinitrotoluene (TNT). It is a detonating explosive.

High-Density Extra Dynamite – See *Dynamite, High-density Extra.*

High-Energy Initiator – Exploding bridge wire systems, slapper detonators, and EEDs with similar energy requirements for initiation.

High Explosive – Any explosive that detonates. In practice, the term is usually confined to explosives that decompose by *detonation.* Hence, high explosives are also called detonating explosives.
A high explosive is characterized by a very high (supersonic) rate of reaction, high-pressure development, and the presence of a detonation wave in the explosive.

In a high explosive, *detonation* takes place, i.e., the chemical reaction propagates at a speed greater than that of sound and a shock wave is generated, supported by energy from the chemical reaction. The reaction must also be accompanied by a shock wave for it to be considered a high explosive. Decomposition by detonation is a very rapid, nearly instantaneous process, so the action of high explosives is fast and violent. The rate of detonation of high explosives ranges from 1,000 to 8,500 meters (3300 to 2800 feet) per second. Although high explosives can be detonated without being enclosed, they are more powerful when confined. A high explosive is more powerful than a *low explosive.*

It is one of the two basic categories of explosives, the other being *low explosives.* High explosives are conventionally subdivided into two classes and differentiated by sensitivity, namely *Primary explosives*, and *Secondary explosives.* Some secondary explosive is so insensitive that it needs a *booster* explosive to initiate detonation. That means the primary sets off the booster, which sets off the secondary. There are 3 kinds of high explosives, namely *Primary*, *Booster* and S*econdary.* High explosives are also categorized into *non-cap sensitive* and *cap sensitive* types.

High explosives are widely used in detonators and blasting charges. In military sector, high explosives are used as bursting charges for bombs, projectiles, grenades, mines and for demolition.

Synonym: Detonating Explosive.

High Explosive Antitank – Shell ammunition for defeat of armor by use of a shaped charge.

Acronym: HEAT.

High Explosive Plastic – Shell with deformable nose, designed to contain a plastic explosive, for use against armor; shock transmitted through the armor causes the back of armor plate to spall.

Acronym: HEP

Synonym: Squash-Head Shell.

High Explosive, Primary – See *Primary High Explosive.*

High Explosive, Secondary – See *Secondary High Explosive.*

High Explosive Shell – Projectile with a bursting charge of high explosive, used against personnel and material.

High Explosive, Tertiary – See *Tertiary High Explosive.*

High Explosives Train – The basic high explosive train consists of the detonator, booster, and bursting charge. High explosives trains may also be either two-step type, such as detonator and dynamite. A typical example of three-step high explosive train consists of the detonator, booster and ANFO.

Also see: *Explosive Train.*

High Melting Explosive – It is one of the names given to the chemical compound Cyclotetramethylenetetranitramine ($C_4H_8N_8O_8$), better known by its acronym *HMX.* It is so named because it is the highest-energy solid explosive. Due to this property, HMX is used exclusively for military purposes.

Also see: *HMX.*

High Order Detonation – A detonation rate equal to or greater than the stable detonation velocity of the explosive.

High Tension Detonator – An early form of detonator that required a voltage exceeding 36 volts to fire it.

HMX – The U.S. acronym for High Melting Explosive, and the UK acronym for Her/His Majesty's Explosive. Chemical name of this Secondary high explosive compound is Cyclotetramethylenetetranitramine, IUPAC being 1,3,5,7-tetranitroperhydro-1,3,5,7-tetrazocine. Chemical formula of the compound is $C_4H_8N_8O_8$. Its molecule is an eight-membered ring of alternating carbon and nitrogen atoms, with a nitro group attached to each nitrogen atom. It is also called *Octahydro-1,3,5,7-Tetranitro-1,3,5,7-Tetrazocine.*

It is a white, crystalline solid with nitrogen content of 7.84%. Molecular weight of this explosive is 296.20 with a melting point of 276 to 286°C. It is slightly soluble in water. Because of its high molecular weight, it is one of the most powerful chemical explosives, although relatively insensitive. Rate of detonation is 9100 meters (29,900 feet) per second with a density of 1.91 g/cm^3.

This nitramine high explosive is chemically related to *RDX*. It is a by-product of RDX manufacture.

Because of the characteristics of HMX, the acronym is also translated High Molecular-weight RDX or even High-velocity Military Explosive.

This explosive is mainly used for military purposes. HMX is used as an explosive charge when desensitized, as a booster charge in admixture with TNT (octol), and as an oxidizer in solid rocket and gun propellants. It is also used in admixture with TNT in high blast warheads.

HMX is used in melt-castable explosives when mixed with TNT, which as a class are referred to as *octol.*

Two grades of HMX are used for military purposes, namely Grade A and Grade B. Grade A has a minimum purity of 93% and a 7% maximum RDX content. Grade B has a minimum purity of 98% and a 2% maximum RDX content.

Synonym: *Octogen.*

HNIW – Acronym for *Hexanitrohexaazaisowurtzitane.*

HNS – Acronym for hexanitrostilbene, $C_6H_2(NO_2)_3$. Its molecular weight is 450.24, nitrogen content is 18.67% and the melting point is 316°C. HNS is a heat resistant explosive, commonly used in deep well charges found in the oil field or in applications requiring the explosive to withstand significant temperatures before initiation. It is relatively insensitive to heat, spark, impact and friction, yet it finds wide use as a heat resistant booster charge for military applications.

Synonym: *Hexanitrodiphenylethylene.*

Hole – As applied to machine explosives, hole means any cavity that is more than one-half diameter deep, being cut by any tool with the direction of feed along the axis of rotation.

Hole, Blast – See *Blast Hole.*

Hole, Block – See *Block Hole.*

Hole, Bore – See *Bore Hole.*

Homemade Bomb – See *IED.*

Homemade Explosive Booby Trap – See *IED.*

Homogeneous Explosion – Homogeneous explosion is one of the two types of *rarified explosions*, the other being heterogeneous explosion. Combustible gas or vapor when mixed with air to form what is known as *explosive mixture* may cause explosion when it comes in contact with a flame or spark. This rarified explosion is a homogeneous explosion.

Compare: *Heterogeneous Explosion,* and *Particulate Explosion.*

Hose, Semi-conductive – See *Semi-conductive Hose.*

Hot Work – See *Work, Hot.*

Hot Zone – See *Zone, Hot.*

Howitzer – A weapon firing slower than a gun and faster, but at lower angle, than a mortar.

Hydrated Explosive – A type of American permissible explosive containing various percentages of salts with water of crystallization, such as Magnesium Sulfate, $MgSO_4$, $7H_2O$.

Hydrazine, Anhydrous – Anhydrous Hydrazine: It is high explosive compound, composed of hydrogen and nitrogen. Its formula is $NH_2.NH_2$. It is a powerful explosive, and because of its explosive ability, it is much used today in rocket fuels.

This material is very toxic and very hazardous to use. It is very sensitive and must not be used without full and complete instructions from the manufacturer as to handling, storage and disposal.

Hydride – A hydride is a combination of hydrogen and a metal. Most hydrides are dangerous materials, powerful reducing agents, and some are toxic. An example of a hydride that is very powerful explosive is boron hydride. The hydrides of boron are used today as rocket propellant fuels. Hydrides can be very sensitive and can explode when exposed to heat, friction, or even vibration.

Hydrogen Bomb – A bomb whose violent explosive power is due to the sudden release of atomic energy resulting from nuclear fusion of hydrogen isotopes. In fusion, light nuclei, such as hydrogen atoms, are joined together at very high temperatures and pressure to form heavier elements, such as helium nuclei, and the end product weighs less than the components forming it. The difference in mass is converted into energy. Since extremely high temperatures are required to initiate fusion reactions, a hydrogen bomb is also known as a thermonuclear bomb.

Lithium deuteride, a compound of lithium and deuterium, is capable to initiate a number of fusion processes involving the hydrogen isotopes, deuterium and tritium. These reactions also produce high-energy neutrons capable of causing fission in a surrounding layer of the most abundant isotope of uranium, ^{238}U so that further energy is released.
The U.S. used the first hydrogen bomb in 1952, and the former Soviet Union used the second one in 1953.

Synonym: *Fusion Bomb, H-bomb,* and *Thermonuclear Bomb.*

Also see: *Nuclear Weapon.*

Hydrox – A steel tube device using low temperature gas produced from a chemical cartridge for producing a blasting effect.

Hypergolic – A two-component propellant system capable of spontaneous ignition upon contact.

IC – Acronym for *Igniter Cord.*

IED – An acronym for Improvised Explosive Device – a term used to describe a homemade bomb or homemade explosive booby trap.

Ignitacord – Ignitacord is a cordlike fuse that burns progressively along its length, with an external flame at the zone of burning, and is used for lighting a series of safety fuses in sequence. It burns with a spitting flame like a sparkler.

Ignite – To start a fuel burning.

Igniter (same as Igniter Cord) – In general any device, chemical, electrical or mechanical, used to ignite something. It is a device that consists of an easy burning composition and is used to amplify the ignition of a propelling charge by a primer. Igniter is an initiator that is used to start deflagration of low explosives. It may also used to amplify the initiation of a primer in the functioning of certain fuses and bursting charges.

Also see: *Electric Match.*

Igniter Charge – Device containing a ready-burning composition, usually black powder, used to amplify the flame from a primer to assist in the ignition of a propelling charge, expelling charge or burster. It is used in certain types of igniting fuses. It is sometimes called an igniter, but is preferably called an igniter charge.

Igniter Cord – A cord for igniting safety fuse. Igniter Cord is used to transmit ignition from a device to a charge or primer. It burns progressively along its length with an external flame.

It may be used as a means of lighting the safety fuse of a plain detonator, or to connect several plain detonators. Its function is to ignite the cords in the desired sequence.

The cord consists of textile yarns covered with black powder or another fast-burning pyrotechnic composition and a flexible protective covering. It may contain a metal core wire or textile threads to improve the strength.

Igniter Cord Connector – An igniter cord connector is used to ensure the safe transmission of the sparking combustion of the igniter cord into the gunpowder core of a connected safety fuse.

Igniter, Dark – Dark Igniter: The dark igniter means priming charge of low luminosity for tracer ammunition.

Igniter, Delay Electric – Delay Electric Igniter: It is a small metal tube containing a wire bridge in contact with a small quantity of ignition compound.

Igniter, Electric – Electric Igniter: It is a primer element for which the receptor is an igniting composition or, in some cases, a detonating composition. Some kinds of electric igniters are called squibs.

Synonym: *Electric Primer* and *Squib.*

Igniter Fuse – The use of a naked flame to ignite the safety fuse is dangerous in a fiery coalmine. Hence devices are made for lighting the fuse in a closed space. These devices are termed igniter fuse.

Igniter, Rocket Motor – See *Rocket Motor Igniter.*

Igniter Safety Mechanism – A device designed and used for arming or safing (aligning or interrupting) an initiation train of an explosive device, such as a rocket motor or gas generator.

Igniter Train – See *Burning Train* and *Explosive Train.*

Ignitibility – The ease with which burning of substance may be initiated.

Igniting Mixture – Chemical mixture used in pyrotechnics.

Igniting Fuse – See *Igniting Fuse.*

Igniting Primer – Primer designed to be initiated by flame from another primer. It may also be used in sub-caliber guns in order to allow drill or practice with the regular primer.

Ignition – The process of starting a fuel mixture burning, or the means for such a process.

Ignition, Behavior on – Behavior on Ignition: Once ignited, an explosive can burn at a controlled rate provided heat and gas are free to escape. It can deflagrate, if the burning rate increases due to thermal conduction and radiation. It can detonate, if the deflagration front reaches shock wave conditions. Shock wave conditions prevail when heat and pressure reinforce the shock front.

Ignition Cartridge – An igniter in cartridge form that can be used alone or with additional propellant increments as a propelling charge for certain mortar ammunition.

Ignition Mechanism – There are four main types of ignition mechanisms. These are friction, heat, impact and spark.

Also see: *Behavior on Ignition.*

Ignition, Means of – Means of Ignition: A device intended to cause deflagration of an explosive (for example: primer for a propelling charge electric squib, igniter for a rocket motor).

Compare: *Initiation, Means of.*

Ignition Point – See *Ignition Temperature.*

Ignition Point, Auto – Auto-ignition Point: The temperature a substance must reach before it will ignite in the absence of a flame. Different substances have different flash points and auto-ignition temperatures.

Ignition Point, Self – Self-ignition Point: Synonym of *Auto-ignition Point.*

Ignition, Spontaneous – Spontaneous Ignition: See *Spontaneous Combustion.*

Ignition System – Arrangement of components provided to initiate combustion of propellant charge.

Ignition System, Leher – Leher Ignition System: Non-electric initiation system based on shock tube technology. It is the latest non-electric initiation system. A shock wave initiated by special initiator or a plain/electric detonator travels through the shock tube at the speed of approximately 2000 meters (6500 feet) per second. The flash of shock wave initiates the detonator assembled at the other end of the tube, which in turn initiates the explosives charge.

Ignition, Through-Bulkhead – See *Through-Bulkhead Ignition.*

Ignition Temperature – The lowest temperature at which combustion begins and continues in a substance when it is heated in air. It is the temperature at which spontaneous ignition of a substance can take place. The combustible substance has to be raised to the ignition temperature, apart from the supply of oxygen, for the fire to occur.

Synonym: Autogenous Ignition Temperature, and Ignition Point.

IHE – Acronym for *Insensitive High Explosive.*

Illuminant Composition – A mixture of materials used in the candle of a pyrotechnic device to produce a high intensity light as its principal function. Materials used consist of an oxidizing agent, a reducing agent (fuel), a binder and a color intensifier plus waterproofing agent. The mixture is loaded in a container under pressure to form the illuminant charge.

Illuminating Ammunition – Illuminating Ammunition is a projectile with a time fuse that sets off a parachute flare at any desired height. It is used for lighting up an area.

Synonym: *Illuminating Shell.*

Illuminating Bomb – Synonym of *Illuminating Ammunition.*

Illuminating Shell – Synonym of Illuminating Ammunition.

IMDG Code – Abbreviation for "International Maritime Dangerous Goods Code". It contains all of the regulations for the transport of dangerous goods by ocean-going ships, inter alia about their classification, packaging and stowing.

IMO – Abbreviation for International Maritime Organization. The organization with its headquarters in London publishes the International Maritime Dangerous Goods (*IMDG Code*).

IME – Acronym for the Institute of Makers of Explosives. It is a non-profit safety-oriented trade association representing leading producers of commercial explosive materials in North America. It deals with the use of explosives, concern with safety and protection of employees, users, the public and the environment in the manufacture, transportation, storage, handling, use and disposal of explosives used in blasting and other essential operations. IME serves as a source of reliable data about the use of explosives and provides technically accurate information and recommendations concerning explosives.

IME Fume Classification – A classification indicating the amount of poisonous or toxic gases produced by an explosive or blasting agent. The IME fume classification is expressed as follows:

Cubic feet poisonous gases per ($1^1/_2$" ×8")

Fume class	Cartridge of explosive material.
1	Less than 0.16
2	0.16 to 0.33
3	0.33 to 0.67.

IME No. 8 Test Detonator – The detonator has 0.40 to 0.45 grams PETN base charge pressed to a specific gravity of 1.4 g/cc and primed with standard weights of primer, depending on manufacturer.

Impact Fuse – Fuse designed to function on impact.

Implosion – An inward burst of particles, fragments, etc., due to reduced pressure, as of a vessel from which the air or vapor has been exhausted.

Antonym: *Explosion.*

Impulse – In rocketry, product of the average thrust in pounds or kilograms by the burning time in seconds.

Impulse Explosive – Burning or low explosive used to propel projectiles from guns, to propel rockets and missiles, launch torpedoes, and launch depth charges from projectiles.

Incendiary – 1. An explosive device or agent designed and intended to cause or set fire. 2. Designates a highly exothermic composition or material that is primarily used to start fires. 3. Filling for incendiary munitions such as shells, bombs, grenades and flamethrowers.

Incendiary Bomb – An incendiary bomb is comprised of a combustible body of magnesium metal, inside of which there is an igniter composition such as thermit. If ignited, the body of an incendiary bomb burns with intense heat.

Incendivity – Incendivity is the property of an igniting agent whereby the agent can cause ignition. Heat, spark, flame, etc. are examples such a property.

Independent – Non-gelatinous permissible explosive used in mining.

Independent Gel – Gelatinous permissible explosive used in mining.

Indirect Contact with Explosive – Bare explosives, the metallic casing of an explosive, or the firing leads of an explosive device make contact with electrical instrument or equipment through electrically conductive equipment or surfaces other than the equipment leads.

Indian Fire – Synonym of *Bengal Fire.*

Induction Time – The time between the breaking of the fuse wire and the detonation of the base charge in firing an electric detonator.

Induction, Toroid – See *Toroid Induction.*

Industrial Explosive – Synonym of *Commercial Explosive.*

Inert – The condition of a device that contains no explosive, pyrotechnic or chemical agent.

Initiate – To initiate means the act of detonating a high explosive by means of a detonator or by detonating cord.

Initiating Device – Devices used to initiate the burning of a propellant or to initiate the reaction of an explosive. The initiation of an explosive reaction requires the application of energy in some form. Propellants are commonly ignited by the application of flame, while explosives are set off by a severe shock.

The device used to initiate the burning of a propellant is called a primer. The device used to initiate the reaction of an explosive is called detonator. Initiating devices are activated by external stimuli sources such as friction, spark, flame, and others.

Depending on their prime source of energy, initiating devices fall into two categories, electric and non-electric.

Depending on their application and external initiation mechanism, initiating devices can be classified into:

- primers (or igniters),
- detonators,
- electric igniters and electric detonators,
- safety fuses,
- detonating cords.

Initiating Explosive – An explosive that serves to initiate the ignition of propellants and the detonation of high explosives. An initiating explosive acts when subjected to heat, impact, or friction. Very small quantity of initiating explosive will detonate on contact with a flame, on mild or low impact or as a result of friction. It is possible to transmit detonation to other explosives close to them.

Initiating explosives may function by themselves, as does the primer cap in a small-arms cartridge. However, in most instances the initiating charge is the lead element in what is known as an explosive train. An explosive train uses the impulse of an initiating explosive to initiate the chain reaction that leads to the detonation of a main burster charge or ignition of a propellant.

Generally explosions of initiating explosives are not as powerful as other explosives. For this reason they are not usually used as explosives themselves. Because of their high sensitivity, they are used to set off explosions in a less sensitive material. Due to their sensitivity, they are usually only kept in small quantities and well away from other explosives.

The sensitive explosives used as initiating explosives are the primary high explosives. Under normal conditions, initiating explosives will not burn, but they will detonate if ignited. They are used in varying amounts in the different primers and detonators and may differ in sensitivity and in the amount of heat given off. Mercury fulminate, lead azide, lead styphnate and diazodinitrophenol (DDNP) are examples of initiating explosives.

The initiation of an explosive reaction requires the application of energy in some form. *Propellants* are commonly ignited by the application of flame, while explosives are set off by a severe shock. The device used to initiate the burning of a propellant is called a *primer*. The device used to initiate the reaction of an explosive is called *detonator*.

Initiation – Usually the term ‘initiation’ means the act of detonating a high explosive by means of a detonator, a mechanical device, or other means.

The term is also used to denote the act of detonating the initiator. It is in effect the first action in the first element of an explosive train.

An initiation system that may serve both as an initiator and as an explosive charge is detonating cord.

Blasting efficiency depends on type of initiation, priming, or boosting.

Also see: *Shock Tube System.*

Initiation, Means of – Means of Initiation – A device intended to cause the detonation of an explosive, such as detonating fuse, detonator, or detonator for ammunition.

Compare: Ignition, Means of.

Initiation of Explosive Reaction – The application of energy initiates an explosive reaction. The preferred method of initiation depends on the characteristics of the individual explosive. The most commonly used methods of initiation are by heat, shock (detonation) and influence.

The application of heat in some form usually initiates low explosives. Initiation by direct blow (percussion) or by friction is simply initiation by heat derived from the energy of these actions.

The sudden application of a strong shock usually initiates the explosive reaction in high explosives. This shock or detonation is created when a smaller charge of a more sensitive high explosive that is in contact with or in close proximity to the main charge explodes. Heat or shock can readily explode the charge.

Detonation of an explosive mass can be transmitted to other masses of high explosive in the vicinity without actual contact. The transmission is caused either by the passage of an explosive percussion wave from one mass to the other or by fragments. The second explosion occurring under these conditions is initiated by influence. Such a phenomenon is termed as *sympathetic detonation* or *sympathetic explosion.*

Initiation Sequence – Synonym of *Firing Train.*

Initiation Stimulus – Initiation stimulus is the energy input to an explosive in a form potentially capable of initiating a rapid decomposition reaction.

Typical initiation stimuli are heat, friction, impact, electrical discharge, and shock.

Initiation, With its Own Means – Ammunition or explosives having their normal initiating device, such as detonators or squibs, assembled to the ammunition or explosives so that the device is considered to present a significant risk of activation during storage or transport.

Initiation, Without its Own Means – Ammunition or explosives that:

(1) Are not stored with an initiating device assembled to the ammunition or explosives; or
(2) Have the initiating device assembled to the ammunition or explosive, but (a) safety features preclude initiation of the explosives filler of the end item in the event of accidental functioning of the initiating device, or (b) the initiating device does not contain any primary explosives and has a high threshold of initiation (e.g., EBW or Slapper detonators).

Initiator – The term 'initiator' is used in the explosive industry to describe any device that can be used to start a detonation or deflagration.

Minute quantity of very sensitive primary high explosive is used to start the detonation of another less sensitive secondary high explosive. Mercury fulminate and lead azide are the main chemicals used in initiators meant for secondary high explosives.

The device used to initiate a high explosive is called a detonator. Detonating cord used to start detonation is also an initiator. Of course, in some initiation systems detonating cord may serve both as an initiator and an explosive charge.

The device that starts a burning or deflagration is called a squib or igniter.

A variety of electrical devices (electric ignitors) were invented in the late 1700s, and used in limited form for the initiation of black powder charges. Alfred Nobel invented a blasting cap in the mid-1860s, consisting of a copper shell filled with mercury fulminate, which was designed for use with safety fuse. The combining of an electric ignitor, a piece of safety fuse, and a detonator resulted in the first delay detonation system. The sensitive mercury fulminate was replaced as the base charge in detonators in the 1930s by *PETN*, a less shock and heat-sensitive material.

In the 1930s modern detonating cord was introduced, although from about 1913 a lead-cased variety was in fairly widespread use in the United States.

Connectors with built-in delay detonators, allowing the design of complex delay shots, became available in the 1950s.

In the mid-1960s the shock tube non-electric detonation system was invented in Sweden, and by the early 1980s its use had spread worldwide.

In the late 1990s electronic detonators with built-in computer chips were in the early stages of introduction worldwide.

The term 'blasting cap' is not usually used today and has been replaced with the term 'detonator'. The term 'shock tube detonator' refers to a device that includes a length of shock tube.

Initiator, High-Energy – High-Energy Initiator: Exploding bridge wire systems, slapper detonators, and EEDs with similar energy requirements for initiation.

Inorganic Explosive – Many inorganic explosives are primary explosives. They are very sensitive to impact and friction, and are easily ignited and grow to detonation from hot spots such as those caused by sparks, flame, and other sources of heat and high temperature. Inorganic compounds do not have hydrocarbon backbones forming the basis of the molecules.

Mercury fulminate, Lead azide, Silver azide, Ammonium nitrate are examples of inorganic explosives. Mercury fulminate was one of the earliest of the initiating explosives. It was extensively used in blasting caps and primers. It has since been displaced by lead azide in modern initiators.

Ammonium nitrate is not a particularly good explosive by itself. It is extremely difficult to initiate, and will propagate a detonation only in very large diameters (above a half-meter).

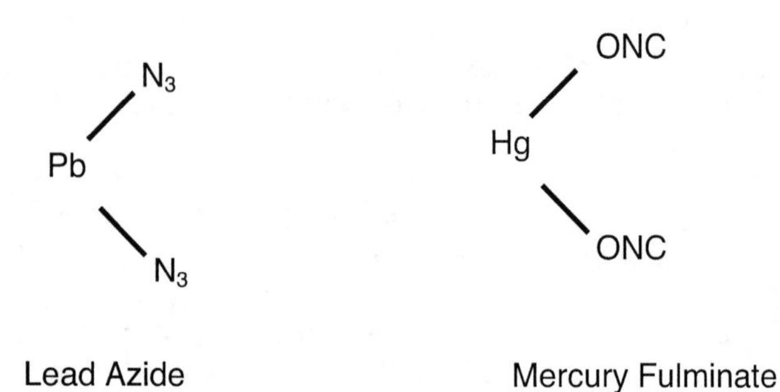

Lead Azide   Mercury Fulminate

Inorganic Nitrate – See *Nitrate, Inorganic.*

In-process Storage Facility – See *In-process Storage Magazine.*

In-process Storage Magazine – See *Service Magazine.*

Synonym: - facility, -vault, -rest house.

In-process Storage Vault – See *Service Magazine.*

Insensitive High Explosive – Explosive substances that, although mass detonating, are so insensitive that there is negligible probability of accidental initiation or transition from burning to detonation.

Acronym: IHE.

Instantaneous Blasting – The practice of initiating all explosive decks, boreholes, or rows of boreholes fired essentially simultaneously.

Instantaneous Detonator – It is a *detonator* that has a firing time of essentially 0 seconds, in comparison to delay detonators with firing times of from several milliseconds to several seconds. It is a detonator that contains no delay element. An instantaneous detonator may be either electric or non-electric.

However, the nomenclature 'instantaneous' is somewhat misleading, as the function times can be as long as 5 milliseconds after the application of the initiating energy. The function time of instantaneous non-electric detonators must include the travel time of the initiating signal through the cord or shock tubing.

Synonym: *Zero Delay Detonator.*

Instantaneous Electric Detonator – The detonator in which the initiation is done via electric current passed through leg wires. It is the substitute of safety fuse with electric wires connected to a fuse head. First prototype instantaneous electric detonator emerged in late 1880s.

Instantaneous Fuse – A fuse, consisting of cotton yarns impregnated with meal powder, which propagates by burning at very high velocity.

Instantaneous Fuse – A fuse that is designed to burst the projectile on the outside of a hard surface, such as a concrete emplacement, before penetration or ricochet. Instantaneous fuse creates some crater on hard ground.

Synonym: Super Quick Fuse.

Institute of Makers of Explosives – See *IME.*

Institute of Makers of Explosives No. 8 Test Detonator – See IME No. 8 Test Detonator.

Integral System – An *LPS* system that has strike termination devices mounted on the structure to be protected. These strike termination devices are connected to the earth electrode system via down conductors. Metallic structure members can serve as parts of the LPS.

Intermediate Charge – Generally, intermediate charge is synonym of *booster.* Still, in some *explosive trains* an intermediate charge functions between the initial charge and the booster to ensure the detonation of the booster.

International Maritime Dangerous Goods Code – See *IMDG.*

International Maritime Organization – See *IMO.*

International Society of Explosives Engineers – The International Society of Explosives Engineers is a non-profit professional society of individuals involved in the explosives industry. The society serves the industry and its members as a clearinghouse for technical information, a voice for the explosives industry to governmental agencies and the general public, a forum for the discussion of explosives issues and technologies, an advocate for the safe and controlled use of explosives, and a bridge between related international organizations.

Acronym: ISEE.

Intrinsically Safe – An apparatus or system whose circuits are incapable of producing any spark or thermal effect capable of causing ignition of a mixture of flammable or combustible material. Intrinsically safe equipment must be tested and approved by an independent agency to assure its safety.

Jell – Something gelatinized, such as jellied gasoline (gasoline thickened with Napalm).

Jet – The central stream of high velocity particles or gases from a shaped charge.

Jet Engine Starter Cartridge – Contrivances used to activate mechanical starters for jet engines. They consist of suitable cases, each containing a pressed block of propellant explosive, and having at its top end a small compartment or recess enclosing an ignition device consisting of an electric igniter and small amounts of black powder or smokeless powder or both.

Jet Fire – A fire from the ignition of a jet of flammable material.

Jet Flame – The flame formed due to the combustion of material emerging with significant momentum from an orifice.

Jet loader – A system for loading ANFO into small boreholes in which the ANFO is drawn from a container by the venturi principle, and blown into the hole at high velocity through a semi-conductive loading hose.

Jet Perforator – Jet perforators are shaped charges designed and used for perforating oil well casings.

Jet Perforating Guns, Oil Well Charged – Oil Well Charged Jet Perforating Gun: Steel tubes or metallic strips into which are inserted *shaped charges* connected by detonating cord.

Jet Tapper – Jet tapper means shaped charge designed and used for tapping open-hearth steel furnaces.

Jet Thrust Unit, including jato – It is used to assist airplanes in taking off, to drive moving targets for practice firing or to propel large missiles. It is constructed of metal cylinders containing a propellant explosive composition capable of burning rapidly and of producing considerable pressure. It is included in the term *rocket motor*.

Joint, Grief – Grief Joint: It is a hollow bar attached to the top of the drill column in rotary drilling.

Synonyms: Kelly Bar, Kelly Joint, and Kelly Stem.

Joint, Kelly – Kelly Joint: See *Joint, Grief.*

Jolt and Jumble Test - Tests intended to simulate the shocks various components of ammunition are subjected to, in transportation and handling.

Kast Test – A German test for brisance of an explosive.

Kelly – It is a hollow bar attached to the top of the drill column in rotary drilling.

Synonym: Grief Joint, Kelly Bar, Kelly Joint, and Kelly Stem.

Kelly Bar – See *Kelly*.

Kelly Joint – Synonym of *Kelly.*

Kelly Stem – See *Kelly*.

Kerosole – Metal dispersion in kerosene.

Kieselguhr – Diatomaceous earth, infusorial earth. A mass of hydrated silica (SiO_2) formed from skeletons of minute plants known as diatoms. It is a very porous and absorbent material, used for filtering and absorbing various liquids, in the manufacture of dynamite and in other industries.

Kiloton Bomb – A nuclear weapon with an explosive power equivalent to one kiloton of T.N.T. (approximately 4×10^{12} joules).

Kiloton Weapon – A nuclear weapon with an explosive power equivalent to one thousand tons of TNT.

Compare: *Megaton Weapon.*

Kinetics of Explosive Reaction – Explosives are comparatively unstable and high-energy content compounds. Energy is released upon explosion to form stable elements or compounds.

An oxidation reaction is the chemical reaction that occurs when a fuel is burning or an explosive is detonating; it is the same in both cases. Oxidation reactions produce heat because the internal energy of the product (final) molecules is lower than the internal energy of the reactant (starting) molecules.

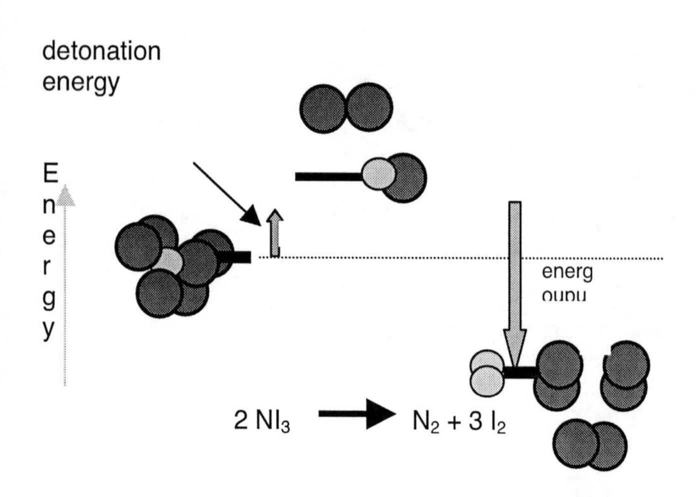

King – Non-gelatinous permissible explosive; used in coal mining.

Kistiakowsky-Wilson Rule – Kistiakowsky and Wilson developed a set of rules during World War II that clarifies the decomposition scenario upon the detonation of HMX ($C_4H_8N_8O_8$) wherein carbon monoxide, carbon dioxide, carbon, water, etc. are formed as follows:

Rule #1 – Carbon atoms are converted to carbon monoxide.
Rule #2 – If any oxygen remains, hydrogen is oxidized to water.
Rule #3 – If any oxygen still remains, carbon monoxide is oxidized to carbon dioxide.
Rule #4 – All the nitrogen is converted to nitrogen gas, N_2.

Knocking – Violent explosions in the cylinder of a petrol engine, often due to over-compression of the mixture of air and petrol vapor ahead of the flame front. The result is a shock wave that causes overheating, plug damage, and loss of power. It is overcome by the use of high-octane fuels containing tetraethyl lead. It can also be avoided by an engine design that increases turbulence in the cylinder head.

Kodaloid™ – Transparent nitrocellulose sheets. Manufactured by Eastman Kodak.

Kodapak™ – Thin, transparent cellulose acetate sheets. Manufactured by Eastman Kodak.

Label – When explosives or other hazardous substances are packaged for transport and storage, a distinctive diamond-shaped label must appear on the package. In the case of explosives, the label has an orange background, with the word 'Explosive' in black letters in the center. A black exploding object appears at the top, and number and division of the explosive, such as 1.4, appear at the bottom of the label.

Lag Time – In firing electric detonators, the interval between the application of the current and the breaking of the bridge wire.

Lamp, Electric Cap – Electric Cap Lamp: The lamp which is fastened on the front of the miner's cap and consists of a flat portable battery that is strapped around the miner's waist, and is connected by an insulated cord to a small electric light and reflector.

Lamp, Safety – See *Safety Lamp.*

L.D. Powder – An approved low-density explosive for use in safety-lamp coalmines.

Lead – An explosive train component that consists of a column of high explosive, usually small in diameter, used to transmit detonation from one detonating component to a succeeding high explosive component. It is generally used to transmit the detonation from a detonator to a booster charge.

Synonym: Explosive Lead.

Lead Azide – Lead azide (PbN_6) is a primary explosive and obviously very sensitive to friction, heat and shock. The velocity of detonation of lead azide is 4100 to 5200 meters (13,400 to 17,000 feet) per second. It is a crystalline substance. Its color varies from colorless to white.

Lead azide is less sensitive than mercury fulminate. The hygroscopicity of lead azide is very low; hence, water does not reduce its impact sensitivity, as is the case with mercury fulminate. This substance has largely replaced mercury fulminate in military ammunition. Lead azide has a high-ignition temperature and is today the most commonly used primary explosive. It is used in detonators and priming compositions. Since lead azide does not react with aluminum, detonator capsules for lead azide are made of aluminum. Lead azide has an excellent shelf life in dry conditions.

Lead Styphnate – It is also known as lead 2,4,6- trinitroresorcinate [$C_6H_3N_3O_9Pb$]. Lead styphnate is a primary explosive and obviously very sensitive to friction, heat and shock. The velocity of detonation is about 5200 meters (17,100 feet) per second.

It is a crystalline solid with a density of 3.06 g/cm^3 at 20^{0}C. Its color varies from orange-yellow to dark brown. It is less sensitive to shock and friction than lead azide. Lead styphnate is particularly sensitive to fire and the discharge of static electricity. Lead styphnate, when dry, can readily ignite by static discharges from the human body. It does not react with metals.

Because of its high metal content, lead styphnate is a weak primary explosive and hence it is not used alone in filling of detonators. Generally it is used in the *ASA* mixtures for detonators. It is used as an initiating explosive in propellant primer.

Leading Line – The line (wire) connecting the electrical power source with the electric detonator circuit.

Synonyms: *Firing line*, *Leading Wire, Lead Line,* and *Lead Wire*.

Leading Wire – Synonym of *Leading Line*.

Lead Line – Synonym of *Leading Line*.

Lead Wire – Synonym of *Leading Line*.

Leakage Resistance – The resistance between the blasting circuit including lead wires and the ground.

LEDC – Abbreviation for Low Energy Detonating Cord. It is a detonating fuse with a core charge too low to enable it to be used reliably for initiating high explosives. It may be used to initiate non-electric blasting caps.

Leher Ignition System – It is the latest non-electric initiation system based on shock tube technology. A shock wave initiated by a special initiator or a plain/electric detonator travels through the shock tube at the speed of approximately 2000 meters (6500 feet) per second. The flash of shock wave initiates the detonator assembled at the other end of the tube, which in turn initiates the explosives charge.

Leg Wire – It means the two single wires or one duplex wire extending out from an electric detonator, which are permanently attached to the electric detonator.

LEL – Abbreviation for *Lower Explosion Limit.*

Antonym: *UEL.*

Letter Bomb – A Letter bomb is an explosive device usually sent via postal service which is set and designed to explode immediately on opening, with the intention of seriously injuring or killing the recipient.

Synonym: *Mail Bomb*, and *Parcel Bomb.*

Light, Bengal – See *Bengal Light.*

Lighter, Fuse – Fuse Lighter: The lighter is a small hollow pasteboard or metal tube with either a wick or two wires connected to a small quantity of igniting composition. It is used for lighting safety fuse, which is inserted at the open end. It may or may not be electrically actuated.

Light, Nacked – See *Nacked Light.*

Lightning Protection System (LPS) – A lightning protection system is a complete system of strike termination devices, conductors, ground terminals, interconnecting conductors, surge suppression devices, and other connectors or fittings required to complete the system.

Lightning Warning System – A system that detects the presence and range of lightning activity and thereby issues an alert or warning.

Linear Shaped Charge – Less flexible, or rigid, version of *flexible linear shaped charge.*

Line, Leading – See *Leading Line.*

Synonyms: *Firing line*, *Leading Wire, Lead Line,* and *Lead Wire.*

Line, Lead – See *Lead Line.*

Line, Firing – See *Firing Line.*

Liquid Carbon Dioxide Explosive – Liquid carbon dioxide is an excellent explosive for use in coalmines. Its low shattering effect is desirable in increasing the proportion of larger sizes of coal. Liquid carbon dioxide is enclosed in hollow steel cylinders that can be shattered by an electric current conveyed to a mild steel disc on the cylinders.

Liquid Oxygen Explosives (LOX) – A high explosive composed of absorbent carbonaceous or some other flammable material, such as cellulose, lamp-black, soot, cork-meal, or industrial earth, saturated with liquefied oxygen, and with or without various metallic powders.

Rarely used today, it has been used in coal mining, quarrying, strip mining, and open-cut ore mining. Its use underground or in confined places is not recommended due to the evolution of a large percentage of carbon monoxide.

It is stronger than black powder. It can be used for some purposes instead of gunpowder and dynamite. It can also be used in mine rescue apparatus and for air-ships.

This explosive has some safety advantages. Since neither of the two components is an explosive at all, there is no risk before the two components are mixed together. Mixing is done at the last moment before firing. On the other hand, there are some disadvantages with made-up explosive. It is very flammable and when it catches fire it will usually detonate. This can be very dangerous to personnel and equipment. Liquid oxygen explosives are not stored as they deteriorate rapidly and lose a great deal of their explosive power in a short time.

The destruction of liquid oxygen explosives is not usually necessary, since they are made up as they are used. However, if enough time elapses after making such an explosive, it will lose much of its explosive power as the liquid oxygen evaporates, and then with proper precautions the remaining material can be disposed. This may be obsolete.

Live Ammunition – Ammunition containing explosives.

Antonyms: *Dummy Ammunition*, and *Drill Ammunition.*

Load, Core – See *Core Load.*

Loader, Air – See *Air Loader.*

Loading, Cast – See *Cast Loading.*

Loading, Melt – See *Melt Loading.*

Lock, Fire – Firelock: A g*unlock* employing a slow match to ignite the powder charge. Also: a gun having such a lock.

Lock, Gun – Gunlock: A gun having a lock that employs a slow match to ignite the powder charge.

Lock, Percussion – Percussion Lock: It means the lock of a gun that is fired by percussion upon fulminating powder.

Long Delay Fuse – See *Fuse, Long Delay.*

Long Period Delay – Delay intervals in the range of 100 milliseconds to 1 second.

Low-density Extra Dynamite – See *Dynamite, Low-density Extra.*

Low-energy EED – All EEDs except exploding bridge wire (EBW) detonators and slapper detonators.

Lower Explosion Limit – The minimum concentration of gas, vapor or dust in air that forms an explosive mixture.

Acronym: *LEL.*

Lower Flammable Limit – The concentration in air of an inflammable material (gas, vapor or dust) below which a burning reaction cannot be sustained.

Acronym: LFL

Low Explosive – A low explosive is one that undergoes a relatively slow chemical transformation, thereby producing a deflagration or an explosion; the speed of the reaction is less than the speed of sound. No shock wave is generated and the reaction is propagated by very rapid burning. A low explosive is characterized by deflagration or a low rate of reaction and the development of low pressure. In order for a low explosive to explode, it must be contained in a strong enclosure. Low explosives burn at a steady speed and are referred to as burning mixtures.

Examples of low explosives are gunpowder, propellants, etc. They are often used as propellants to force a bullet out of a gun or send a rocket into space. Low explosives are suitable for use in igniter trains and certain types of propellants.

Also see: *Burning* and *detonation.*

Low-freezing Dynamite – See Dynamite, Low-freezing.

Low Level Fireworks – Fireworks that ignite and fire their product into the sky directly from the ground.

Low Order Detonation – A chemical reaction in a detonable material in which the reaction front advances with a velocity appreciably lower than the characteristic detonation velocity for the material in question.

LVD – Acronym for *Low Velocity Dynamite.*

Low Velocity Dynamite – Non-nitroglycerin military dynamite, the cartridge material of which is or can be packaged on standard dynamite machinery.

Acronym: LVD.

LOX – Acronym for *Liquid Oxygen Explosive.*

LSC – Abbreviation for *Linear Shaped Charge.*

Lyddite – An explosive consisting of Trinitrophenol [Picric acid, C_6H_2OH-$(NO_2)_3$], mixed with 10% Nitrobenzene and 3% Vaseline.

M-80 – A class of large firecrackers. It was originally developed from a small red cardboard tube approximately 3.75cm (1.5 inches) long and 1.4 cm (0.5625 inches) in diameter, with a fuse coming out of the side, by the U.S. military to simulate gunfire. It has also been manufactured and sold to the public as fireworks.

Machine, Blasting – See *Blasting Machine.*

Machine, Blasting, Capacitor-discharge – See *Capacitor-discharge Blasting Machine.*

Mach Wave – See *Wave, Mach.*

Magazine – 1. Any building, structure, or container, other than an explosives manufacturing building, approved and used for the storage of explosives. Usually it is bullet and theft resistant.

2. A device for ammunition feeding, which holds the cartridges just prior to them being put in the chamber of the firearm by the operation of a mechanism on the firearm. The mechanism may be operated manually as in a bolt action, or semi-automatically when the gun fires after pulling the trigger. The magazine may be detachable or part of the gun.

Also see: *Day Box, Rest House, Storehouse,* and the individual entries hereunder.

Magazine Distance – The minimum distance permitted between any two storage magazines. Both the type of magazine, and the class and quantity of explosives it contains, determine the magazine distance.

Magazine, Field – Field Magazine: Synonym of *Portable Magazine.*

Magazine, In-process Storage – See *In-process Storage Magazine.*

Magazine, Permanent – Permanent Magazine: A magazine that is fastened to a foundation and which is specially designed, constructed and designated for the long-term storage of explosives or ammunition.

Magazine, Pit – Pit Magazine: An improvised magazine for storing explosives temporarily. This magazine consists of an underground damp-proof pit/cavity, the floor of which is either cemented or covered wooden planks. Essentially the explosives in the pit must be covered with tarpaulin awning so as to be protected against sun and rain, and the margin of the pit is raised so as to prevent rainwater from entering. The storage area should be fenced off to prevent unauthorized persons entering the premises. Not commonly used.

Magazine, Portable – See *Portable Magazine.*

Magazine, Service – Service Magazine: An auxiliary building or suitable designated room (vault) designed and used for the intermediate storage of explosives, not exceeding the minimum amount necessary for safe and efficient operation.

Synonyms: In-process Storage Magazine, Rest House, and In-process Storage Vault.

Also see: *Day Box.*

Magazine, Storage – Storage Magazine: A structure designed and specifically designated for the long-term storage of explosives or ammunition. The magazine is usually fastened to a foundation.

Also see: *Storehouse.*

Magazine, Surface – Surface Magazine: A specially designed and constructed structure for the storage of explosive materials on the surface of the ground.

Magazine, Underground – Underground Magazine: A specially designed and constructed structure for the storage of explosive materials underground.

Magnesium Dust – Fine powder of magnesium metal; used in flash light powder, pyrotechnics.

Magnesium Flare – Magnesium flare is used in aviation for signaling and for illuminating landing fields and military objectives. It is composed of mixtures of potassium perchlorate or barium and strontium nitrate with magnesium powder or flakes.

Also see: Aluminum Flare.

Mail Bomb – See *Letter Bomb.*

Synonyms: *Letter Bomb*, and *Parcel Bomb.*

Main Explosive Charge – The explosive that performs the major work of blasting.

Major Accident – *An occurrence including in particular major emission, fire or explosion involving one or more hazardous chemicals and resulting from uncontrolled developments in the course of an industrial activity or due to natural events leading to serious effects both immediate or delayed, inside or outside the installation likely to cause substantial loss of life and property including adverse effects on the environment. [*Quoted from the model rules on 'Control of industrial major accident hazards', Government of India.]

Major Accident Hazard – *Major accident hazard means an occurrence resulting from uncontrolled developments in the course of an industrial activity leading to a serious danger to persons or the environment. [*Quoted from the CIMAH - Control of Industrial Major Accident Hazards Regulation 2006.]

Major Hazard – Major hazard means a large-scale chemical hazard, especially one that may result from an acute event.

Mannitol Hexanitrate – A very sensitive high explosive. It is used as a substitute for fulminates in priming compositions and percussion caps. Mannitol hexanitrate is also used in combination with tetracine for the explosive charges of detonating rivets.

Manometric Bomb – It is a device that consists of a specially designed closed vessel, a pressure-resistant hollow steel body that can be bolted together and has a hole to accommodate a piezoelectric pressure transducer. The main parts of the Manometric bomb are the combustion chamber, chamber closing elements, elements for the electric ignition of the sample, pressure discharge valve, and temperature and pressure gauges.

It is used to study the burn-up properties of a gunpowder or propellant charge powder. The determination of the combustion pressure of propellants (or pyrotechnic compositions) at constant volume conditions is performed in a ballistic bomb. The vessel has a combustion chamber with a volume up to 2.5dm^3, and it can withstand the dynamic pressure of as much as 5000 kgs/cm^2 (73000 psi).

Synonyms: Ballistic Bomb, Bichel bomb, Closed Bomb, and Pressure Bomb.

Manufacturing Code – Code markings stamped on the package of explosives, indicating the date of manufacture and other information.

Maroon – Exploding fireworks used as a warning signal.

Mass Detonation – An explosive mass detonates when a unit or any part of a larger quantity of explosive explodes and causes all or a substantial part of the remaining material to detonate or explode simultaneously. With regard to detonators, 'a substantial part' means at least 90%.

Synonym: *Mass Explosion.*

Mass Explosion – See *Explosion, Mass.*

Mast System – An LPS system that consists of one or more poles with a strike termination device connected to an earth electrode system by down conductors. In the case of a metallic pole, the pole could serve as strike termination device and down conductor. Its purpose is to intercept lightning flashes from the protected area.

Mat, Blasting – See *Blasting Mat.*

Match – 1. Practical method of producing flame. Generally a short splendor piece of wood or other rigid inflammable material tipped or treated with igniting material that bursts into flame under the influence of spark, contact with certain chemicals, or through friction. 2. Cotton wicking impregnated with gunpowder, used for conveying fire to a piece or parts of pyrotechnical devices.

Match, Black – Black match: It is a bare or uncovered match.

Match, Electric – Electric Match: A device designed and used for the electrical ignition of fireworks and pyrotechnic articles that contains a small amount of pyrotechnic material, which ignites when a specified electric current flows through the leads.

Synonym: Igniter.

Match, Paper – Paper Match: A paper match that strikes only on a surface specially prepared for it to ignite.

Match, Percussion – Percussion Match: A match that ignites by percussion.

Match, Quick – Quick Match: It consists of cotton wicking impregnated with gunpowder and covered with a loose paper piping.

It is used for conveying fire to the combustible portion of pyrotechnical devices. The effect of quick match is almost instantaneous.

Quick match is distinguished from a fuse by the fact that its effect is almost instantaneous, whereas a fuse burns at a comparatively slow and exact rate.

Match, Safety – See *Safety Match.*

Match, SAW – Match SAW: Abbreviation of *Strike Anywhere Match.*

-Match, Slow – Slow-Match: It is used when a long delay is necessary. Slow-match smolders away at the rate of 30 cm (1 foot) in several hours. It is made by boiling loose hemp cords in a dilute solution of saltpeter.

Match, Strike Anywhere – Strike Anywhere Match (SAW): The major reactants of the tip of the SAW match are potassium chlorate and phosphorus sesquisulfide; the grit being powdered glass. It operates by thermal ignition. Friction raises the temperature of the fine grits that are included in the match head. These grits, as hotspots, then cause the initiation of burning of the reactants.

Acronym: SAW Match.

Synonym: Wooden Kitchen Match.

Match, Wooden Kitchen – Synonym of *Strike Anywhere Match.*

Maximum Credible Event – The maximum credible event (MCE) from a hypothesized accidental explosion or fire is the worst single event that is likely to occur from a given quantity and disposition of explosives or explosive devices. The event must be realistic with a reasonable probability of occurrence considering the explosive propagation, burning rate characteristics, and physical protection given to the items involved.

Maximum Recommended Firing Current – The highest recommended electric current to ensure safe and effective performance of an electric detonator.

Maximum Sky Brightness – The worst possible sky condition, usually uniform clouds or overcast, for observing pyrotechnic signals.

MCE – Acronym for *Maximum Credible Event.*

MDF – Acronym for *Mild Detonating Fuse.*

Means of Ignition – A device intended to cause deflagration of an explosive (for example: primer for a propelling charge electric squib, igniter for a rocket motor).

Compare: *Means of Initiation.*

Means of Initiation – A device intended to cause the detonation of an explosive, such as detonating fuse, detonator, detonator for ammunition.

Compare: Means of Ignition.

Medicinal Explosive – A very useful heart drug is made up of *nitroglycerin.* This chemical is the active ingredient of dynamite. It is used medically as a vasodilator. The low concentration of nitroglycerin in medication, and the moist environment in the body, make it impossible for it to explode. The nitroglycerin absorbed by the body dilates the blood vessels, increases the blood supply to the heart, and reduces the workload of sending blood around a body. Angina pain is relieved as the heart muscle receives the oxygenated blood it needs for good functioning.
Nitroglycerin also used to be used as a headache remedy under the name 'glonoin' before the invention of dynamite by Alfred Nobel.

RDX, the most powerful high explosive, is used as a medicine under the name of methenamine, for the control of urinary tract infections.

Medium Delay Fuse – A type of delay fuse in which the fuse action is delayed for a period of time between that of short delay and long delay fuses. The delay time is usually four to fifteen seconds.

Medium Velocity Dynamite – Non-nitroglycerin military dynamite, the cartridge material of which is or can be packaged on standard dynamite machinery.

Acronym: MVD.

Megaton Weapon – See *Weapon, Megaton.*

Compare: *Kiloton Weapon.*

Melt Loading – Process of loading an explosive device by melting the explosive and allowing it to solidify in the device.

Mercuric Cyanate – Synonym of *Mercury Fulminate.*

Mercury Azide – Mercury azide, $Hg_2(N_3)_2$ is a primary high explosive. It is a white crystalline solid. It decomposes violently, i.e., explodes on exposure to shock, light, heat.

Mercury azide is used in detonators.

Also see: *Azides.*

Mercury Fulminate – Mercury fulminate, $C_2N_2O_2Hg$, or $Hg(C{\equiv}NO)_2$ is a primary explosive and obviously very sensitive to friction, heat and shock. Its velocity of detonation is 3,500 to 6,430 meters (11,500 to 21,100 feet) per second.

It is a crystalline solid with a density of 4.42 g/cm^3 at 20°C. Its colors may be gray, pale brown, or white. Mercury fulminate is perhaps one of the oldest known initiating explosives. This primary explosive is so sensitive that even the action of dropping a crystal of the fulminate causes it to explode.

It was used in detonators, priming compositions and percussion caps. Its military uses have been taken over to a large extent by *lead azide* because of the poor stability of mercury fulminate at elevated temperatures.

Synonyms: Fulminate of Mercury, Fulminating Mercury, and Mercuric Cyanate.

Also see: *Blasting Cap*, *Fulminate*, and *Percussion Cap.*

Metal Clad Detonating Cord – It consists of a core of detonating explosive clad in a soft metal tube with or without a protective covering.

Metal Clad Detonating Fuse – Synonym of *Metal Clad Detonating Cord.*

Metal Clad Linear Flexible Shaped Charge – It consists of a V-shaped core of detonating explosive and is clad in a flexible metal sheath.

Metallized Explosive – An explosive which is sensitized or boostered with finely divided metal flakes, granules or powders, usually aluminum or ferrosilicon, with the purpose of yielding more energy.

Microballoons –Tiny hollow spheres of glass or plastic that are added to explosive materials to enhance sensitivity by assuring an adequate content of entrapped air.

Mild Detonating Fuse – A flexible metal tube, usually lead, containing a much smaller core of high explosive than the normal detonating cord. More accurately it may be called miniature detonating fuse.

Acronym: MDF.

Synonym: Miniature detonating fuse.

Military Bomb – Military bombs are generally mass-produced weapons, developed and constructed to a standard design out of standard components in factories and intended to be deployed in a standard way each time. The usual method of delivering military bombs to their target is by bombing - dropping them from a bomber airplane. Modern bombs may be guided after they leave a plane by remote control or mounted on a guided missile.

Military Explosive – Explosives have found an application for the purpose of strengthening man's arm in war. All explosives are suitable for use as military explosive. Generally, military explosives need to be stable, thus implying resistance to shock, moisture and other considerations that come about through the nature of their use. They need to be capable of being stored for long periods of time without significant deterioration.

For military purposes the most important are trinitrotoluene (TNT), ammonium picrate, picric acid, nitrostarch, tetryl, amatol, cyclonite, and

PETN. Some of these are not used today (ammonium picrate, nitrostarch, tetryl, etc.).

Military Guncotton – Synonym of *Cellulose Nitrate.*

Mill Cake – A product, in black powder manufacture, taken from the edge runner mills.

Millisecond Delay Detonator – A short delay detonator having built-in-time delays of various lengths. The interval between the delays at the lower end of the series is usually 25 milliseconds. The interval between delays at the upper end of the series may be 100 to 300 milliseconds. The millisecond delay blasting is used to improve rock fragmentation and displacement, provide greater control of blasting vibrations, decrease blast noise and rock throw, reduce powder factors, and reduce blasting costs.

Mine – An encased explosive or chemical charge designed to be placed in position so that it detonates when its target touches it or moves in the vicinity.

Mine with Bursting Charge – The mine that is usually metal or composition container filled with a secondary detonating explosive, designed to be operated by the passage of ships, vehicles or personnel.

Miniature Detonating Fuse – Synonym of *Mild Detonating Fuse.*

Miniaturized Detonating Cord – Detonating cord with a core load of five or less grains of explosive per foot (30 centimeters).

Minimum Recommended Firing Current – The lowest recommended electric current to ensure reliable performance of an electric detonator.

Minimum Gap Sensitivity – An air gap, measured in inches, that determines whether the explosive material is within specific tolerances for gap sensitivity.

Mining, Pillar and Stall – See *Pillar and Stall Mining.*

Misfire – An explosive charge, which for any reason has failed to *fire* as planned. An occurrence where testing before firing a shot reveals broken continuity which cannot be rectified is also included in the definition of misfire. Misfire may be contrasted with hang fire, which is delay in any part of a firing charge.

A misfire is dangerous. Misfires must be treated with care and caution. It is usually difficult and dangerous to resolve misfires. Misfires have to be handled under the direction of the person in charge of the blasting.

If there is any misfire, no attempt should be made to extract explosives from any charged or misfired hole. The hole needs to be reblasted by putting a new primer. Since refiring of the misfired hole presents a hazard, the explosives may be removed by washing out with water or, where the misfire is under water, blown out with air.

In the event of misfire while using cap and fuse, all employees should remain away from the charge for at least one hour. All wires shall be carefully traced and a search made for unexploded charges.

Missile – Any object dropped, fired, launched, thrown, or otherwise projected with the object of striking a target. 'Ballistic missile', or 'guided missile' is usually shortened as missile. Such a missile is projected by a release of energy.

Mix, Bulk – See *Bulk Mix.*

Mix Delivery Equipment, Bulk – See *Bulk Mix Delivery Equipment.*

Mixture, Burning – See *Burning Mixture.*

Mixture, Explosive – Explosive Mixture: 1. Synonym of *Composite Explosive*. 2. A mixture of inflammable gas with air in certain proportion that explodes when ignited.

Also see: *Explosive limit.*

Mock Explosive – The Substance that bears similar physical properties, such as cohesion, density, texture, etc., to an explosive material. It is non-detonable. However, some mock explosives are exothermic materials and can burn. It is used to represent explosives for purposes such as dry run testing of equipment.

Synonym: Mock HE.

Moderated – Implies the presence, in propellants, of a surface coating to the grain that slows the initial rate of burning.

Moderator – A material that reduces explosive sensitivity.

Molotov Cocktail – A crude hand grenade or firebomb made of a bottle filled with inflammable liquid, such as gasoline, and fitted with a wick saturated rag taped to the bottom and ignited at the moment of hurling. The Molotov cocktail was first used by Russians against German tanks in World War II. It is now almost exclusively used by terrorists. In fact, Molotov cocktail is now the generic term for a variety of crude *incendiary weapons*. It was named after the Russian statesman Vyacheslay M Molotov (born 1890).

Synonyms: Benzine Torch, Molotov Bomb, Molotov Grenade, and Petrol Bomb.

Monomethylaminenitrate – A compound used to sensitize some *water gels*.

Mono-oil – It is a mixture of about 62 percent ortho-, 34 percent para-, and 4 percent meta-nitrotoluenes. It is yellow oil with pungent odor. Its specific gravity is 1.16.

Mono-oil is used in low-freezing dynamites. It is also used as plasticizer for nitrocellulose. In blasting explosives it is used in combination with chlorates or nitrates. Such plastic explosives have the desirable property of completely filling the borehole by using only a light pressure. The oily nitro compound also serves as a protection against dampness, and it has the advantage of being an active ingredient in comparison with paraffins and greases which have previously been used as plasticizers.

Monopropellant – A single-phase liquid propellant that contains an oxidizing agent and fuel (combustible substance).

Mortar, Ballistic – Ballistic Mortar: The term means an instrument used in a laboratory for measuring the relative power or strength of explosive materials.

Motor – In relation to rocket, motor is a generic term for a solid propellant rocket consisting of the assembled propellant, case, ignition system, nozzle and appurtenances.

MS Connector – Non-electric, short-interval (millisecond) delay devices for use in delaying blasts that are initiated by detonating cord.

Mudcap – See *Adobe Charge*.

Mud-capping – The blasting of boulders by placing a quantity of explosives against a boulder, rock, or other object without confining the explosives in a drill hole.

Synonym: Bulldozing, Adobe Blasting, and Dobying.

Multi-perforated Powder – In a propellant powder if the grains of the powder are made so that each grain has multiple holes in it, it can burn progressively, meaning that they burn faster as they are consumed. This allows powders to develop more gases after a projectile has begun to move down the barrel of a gun. The powder is then known as multi-perforated powder. It is worth mentioning that propellants can be made to perform differently by varying the physical characteristics of the individual grains.

Also see: *Perforated Powder.*

Multiple Grain – An assembly of solid propellant grains inside an explosive device or motor.

Multi Section Charge – Propelling charge in separate-loading or semi-fixed ammunition that is loaded into a number of powder bags. Range adjustments can be made by increasing or reducing the number of bags used, as contrasted with a single-section charge in which the size of the charge cannot be changed.

Munroe Effect – The concentration of explosive action through the use of a shaped charge.

Musket – An early small firearm used to be fired from the shoulder.

Muzzle Blast – Sudden air pressure exerted in the vicinity of the muzzle of a weapon by the rush of hot gases and air on firing.

MVD – Acronym for *Medium Velocity Dynamite*.

Nail Bomb – An anti-personnel explosive device packed with nails to increase its destructive power. The nails act as shrapnel, leading almost certainly to greater loss of life and injury in inhabited areas than the explosives alone would.

Naked Light – Open flames or fires, exposed incandescent materials or any other unconfined source of ignition.

Napalm Bomb – Now subject to an international ban from use, this jellied fuel is made by the addition of aluminum soap powder of *na*phthene and *palm*itate (hence na-palm) to gasoline, and has been used as a weapon in war. Especially used in incendiary bombs and flamethrowers.

Since gasoline burns so quickly, it is not very effective in war for igniting the target of the flamethrowers. A way to jelly gasoline was discovered that worked quite well. Mixing an aluminum soap powder with gasoline produced brownish sticky syrup that burned more slowly than raw gasoline, and was much more effective at igniting one's target. The napalm was mixed in varying concentrations of 6% (for flame throwers) and 12-15% for bombs mixed on site (for use in perimeter defense). Napalm was extensively used in World War II in flamethrowers and firebombs in the latter part of the war. Napalm bomb was used for aerial bombardment in the Korean War and in Vietnam.

National Fire Protection Association – It is an industry-government association that publishes standards for, among others, explosive material and ammonium nitrate.

Natural Barricade – The expression 'natural barricade' means natural features of the ground, such as hills, or timber of sufficient density that the surrounding exposures, which require protection cannot be seen from the *magazine* when the trees are bare of leaves.

NC – Acronym for *Nitrocellulose.*

NE – Acronym for *Nuclear Explosive.*

Needle, Blasting – See *Blasting needle.*

NEO – Acronym for *Nuclear Explosive Operation.*

Neutron Bomb – See *Nuclear Weapon.*

NFPA – Acronym for *National Fire Protection Association.*

NG – Acronym for *Nitroglycerin.*

Niter – See *Nitre.*

Nitramine Explosive – Judging from the molecular groupings, Nitramines (-NH-NO_2) are a class explosives. In other words, the nitramines are a class of organic nitrate explosives. RDX (hexahydro-1,3,5-trinitro-1,3,5triazine), HMX(octahydro-1,3,5,7-tetranitro-1,3,5,7 tetrazocine), nitroguanidine, and tetryl are significant nitramine explosives. Due to the higher density, and bigger molecules nitramine explosives are relatively powerful.

Also see: Classification of Explosives, and *Explosophore Group.*

Nitramon – Trade name for an explosive containing around 92 percent ammonium nitrate and no nitroglycerine introduced by the DuPont Co. in 1935. It is very insensitive and characterized by an unusually high degree of safety. Its detonation requires a high-velocity primer of TNT or dynamite in addition to the initiator. It is used for blasting in wet and dry work.

Nitrate – Nitrate means salt of nitric acid, e.g., sodium nitrate, $NaNO_3$.

Nitrate, Ammonium – Ammonium Nitrate: Ammonium Nitrate, NH_4NO_3 is a colorless crystalline salt. It is an oxidizing substance. It is mainly used as a fertilizer. Although Ammonium Nitrate in the pure state is very insensitive, it has the record of spontaneous detonation. An impressive but tragic demonstration of the destructive capability of it occurred in 1948 in a ship in Texas City, USA where a shipload of ammonium nitrate blew up, leveling the town and killing five hundred people. Subsequently it has become the most commonly used oxidizer in explosives and blasting agents. It is a very insensitive and stable high explosive. Although it is not regarded as explosive, some categorize it as a *tertiary high explosive.* It is also used as a constituent of secondary high explosives, such as amatol, slurry explosives. It is much used as a constituent of *ANFO*. It is also the chief component of ammonia dynamite. Original *safety explosives* contained ammonium nitrate. It has little use as a military explosive in its simpler forms, but when mixed with other explosives like *TNT* it is encountered frequently in military explosives.

Nitrated Cotton – See *Nitrocellulose.*

Nitrate Explosive, Ammonium – Ammonium Nitrate Explosive: Ammonium Nitrate Explosives are mixtures of ammonium nitrate and carbon carriers, such as wood meal, fuel oils or coal. Sensitizers, such as Nitroglycol or TNT and Dinitrotoluene are added to the mixtures. Aluminum Powder may also be added to improve the strength. These types of mixtures can be cap-sensitive. The non-cap-sensitive mixtures are blasting agents. Many uses of ammonium nitrate as an explosive are facilitated by the addition of about 6% intermediate flash point fuel oil which, among other things, helps keep it dry.

Many permissible explosives are ammonium nitrate in powder form or gelatinous explosives to which inert salts, such as rock salt or potassium chloride is added to reduce their explosion temperature.

The resistance to moisture of powder-form ammonium nitrate explosives may be improved by addition of hydrophobic agents, such as calcium stearate. The densities of the powders are about 0.9 - 1.05 g/cm^3.

Compare: *ANFO.*

Nitrate Mixture Explosive – Nitrate mixture explosive is a class of explosives in which a nitrate is mixed with a fuel. For the purpose of the Explosive Act and the rules thereunder, the term 'nitrate-mixture' means any preparation, other than gunpowder, which is formed by the mechanical mixture of a nitrate with any form of carbon; or with any carbonaceous substances not possessed of explosive properties, whether sulfur is added or not to such preparation, and whether such preparation is not mechanically mixed with any other non-explosive substance, and includes any explosive containing a perchlorate.

Nitrate, Inorganic – Inorganic Nitrate: It is a compound of metal that is combined with the monovalent $-NO_3$ radical. Nitrates are powerful oxidizing agents and have associated with them certain toxicity. The nitrates under certain conditions become explosive, and all of the inorganic nitrates act as oxygen carriers, which under proper conditions can give up their oxygen to other materials, which may in turn detonate. For example, potassium or barium nitrate are added to double-base powders for the purpose of reducing flash and rendering the powder more ignitable. A further use for this material is to mix with a smokeless powder, which is not completely colloided, for the purpose of granulation. An example of such a powder is "E.C. Powder" which is used for loading blank cartridges and hand grenades. Sodium and potassium nitrate are also used in black powder as the oxygen carrier to support the combustion of the sulfur and the charcoal. Ammonium nitrate, however, has all of the properties of the other nitrates, but it is also able to detonate by itself under certain conditions. Ammonium nitrate is therefore a high explosive, although very insensitive to impact and difficult to detonate.

Nitrate, Organic – Organic nitrates are usually termed nitro compounds. However, in almost every case they are nitric acid esters of an organic material. These compounds are a combination of the nitro ($-NO_2$) group and an organic radical. In general, these nitro compounds, or organic nitrates, are flammable and explosive. They can be detonated by means of boosters, by means of shock, temperature, friction or a combination of any of these methods.

Nitre – Nitre is the synonym of *Potassium Nitrate* KNO_3, commonly known as *Saltpeter.* It is the basic ingredient of black powder.

Synonym: *Niter,* and *Saltpeter.*

Nitrocellulose – Chemically, nitrocellulose is a nitric ester of *cellulose* and not a nitro compound, having the empirical formula $[C_6H_7O_2\ (ONO_2)_3]x$. The correct chemical name, therefore, is cellulose nitrate. However, the term nitrocellulose has been adopted by tradition and is generally used.

The first great advance in the chemistry of explosives took place in 1846 when Professor Christian Friedrich Schönbein of the University of Basle invented guncotton, a nitro-cellulose with more than 12.3% nitrogen.

Nitrocelluloses are produced by nitrating cellulose (wood or cotton). This explosive substance is obtained from cellulose by the action of nitric acid in the presence of sulphuric acid upon cellulose. Since the exact molecular weight of cellulose is unknown and since degradation of the cellulose molecule takes place during the reaction, it is not possible to assign absolute formulas to cellulose derivatives. In fact, nitrocellulose is not a single definite compound, but a mixture, the degree of nitration being measured by the percentage of nitrogen. The composition and properties of the cellulose compounds depend on the strength of the acid mixture and the temperature and duration of reaction. Present day custom is to designate nitrocellulose by percent nitrogen rather than by chemical terms, or to use commercial terms that refer to nitrocelluloses of certain percentages of nitrogen.

According to the degree of nitration and use, they are called *guncotton*, *nitrated cotton*, *nitro cotton*, *collodion cotton*, *pyroxylin*, etc. These are the commercial products of interest in the field of explosives.

If ignited unconfined, even the highly nitrated celluloses burn quietly. Since this material is a poor heat conductor and may allow the heat to accumulate, the danger of explosion increases with the quantity burned.

Although the highly nitrated celluloses are high explosives, these are not suitable for use in their natural light fluffy form. However, they can be compressed by suitable measures to a density as high as 1.4. They are then used as the main charge for submarine depth bombs.

Alfred Nobel first gelatinized cellulose nitrate with nitroglycerin and introduced the first double-base smokeless powder. In gelatinized form cellulose is used as the basis of all propellants, particularly cannon powder or in mixture with guncotton in smaller caliber guns; or mixed with nitroglycerin in double-base powders.

Nitrocellulose can be detonated readily with all common detonators. Uncolloided forms of cellulose nitrate are used as a flame carrier in the central tube of shrapnel shell to connect the fuse with the base charge of black powder. Smokeless gunpowder and most of the violent propellants contain some grade of nitrocellulose. Nitrocellulose was the first reliable and stable gunpowder. It is the principal type of propellant today. One of the chief uses for nitrocellulose is in the manufacture of smokeless powder.

Nitrocellulose is the universal basis of propellant powders and hence it is of tremendous military importance.

When nitrocellulose is stored wet or in solution in alcohol and contained in airtight drums is not regarded as an explosive. It is then regarded as a combustible material, a hazardous substance of class 4. It is regarded as a hazardous substance of class 3.2, i.e. a volatile flammable liquid if it is in solution in inflammable liquids, or wetted with more than 40% inflammable liquids by weight.

Nitrocellulose, Modified – Modified Nitrocellulose: It means the gelatinized or plasticized nitrocellulose. As a result of appropriate processing, it loses its natural fibrous structure to become plastic or elastic. It may be in granular form, in flakes, in chips, in blocks or in more or less viscous pastes.

Nitrocellulose Powder – Synonym of *Smokeless Gunpowder.*

Also see: *Gunpowder.*

Nitro Compound – The chemical compound containing the $-NO_2^-$ radical. Nitro Compounds are a class of explosives. Nitro compound explosive is the standard against which all other explosives are measured. The reason that nitro the group ($-NO_2$) leads to unstable compounds is that nitrogen has a charge of +1, and nitro groups have a strong tendency to withdraw (pull) electrons from other parts of the compound. Attaching three nitro groups to a compound leads to an extremely unstable situation, which is the characteristic of a chemical to be an explosive.

Nitro-compound Explosive – Nitro-compound is a class of explosives. It means any chemical compound which is possessed of explosive properties or is capable of combining with metals to form an explosive compound, and is produced by the chemical action of nitric acid (whether mixed or not, with sulphuric acid), or of a nitrate mixed with sulphuric acid, upon any carbonaceous substance, whether such compound is mechanically mixed with other substances or not.

Nitro-compound, Small Arm – See *Small Arm Nitro-Compound.*

Nitro Derivative, Deflagrating Metal Salt – See *Deflagrating Metal Salts of Aromatic Nitro Derivatives.*

Nitrocotton – Synonym of *Colloidion Cotton.*

Also see: Cellulose Nitrate, and *Nitrocellulose.*

Nitrogen Trichloride – An oily or crystalline high explosive. Its color is yellow, and detonates at 95°C. In the dry state it is very hazardous. It must be kept in very small quantities, preferably under water. It is particularly sensitive if it is not pure.

Nitroglycerin – Nitroglycerin, represented by the empirical formula $C_3H_5N_3O_9$, and by the chemical formula $C_3H_5(ONO_2)_3$, is a heavy yellowish poisonous oily explosive liquid. Its specific gravity is 1.6, and is soluble in alcohol and ether.

Nitroglycerine was discovered by the Italian Scientist Ascanio Sobrero in 1846. Twenty years after its invention, *Alfred Nobel* harnessed the explosive energy of nitroglycerin by absorbing it in kieselguhr.

Nitroglycerin is an extremely unstable and terrible explosive. Contrary to some opinions, it is the most powerful of all explosives known. Some might have more brisance (shattering power), but 'nitro' is the standard against which all other explosives are measured. Nitroglycerin explodes on only a four-centimeter drop of a two-kilogram weight. Small quantities of it can readily be detonated by a hammer blow on a hard surface, especially when it is absorbed in filter paper.

Nitroglycerin is an excellent solvent, and because of this property it can be mixed with a wide variety of other materials to adapt it to different conditions. It is one of the most important and most frequently used components of explosive materials.

Nitroglycerin is the active base or ingredient of various types of explosives, such as dynamites or blasting gelatin. A small quantity of nitroglycerin is used in certain classes of explosives, where the nitroglycerin helps to increase the sensitivity of the explosive mixture and to ensure propagation of the explosion. In combination with nitrocellulose and stabilizers, it is the principal component of powders and solid rocket propellants.

The nitroglycerine is also thickened or gelatinized by the addition of a small percentage of nitrocellulose. This process assists in preventing exuding or settling of the absorbent material. In order to prevent settling non-gelled dynamites, boxes of stored non-gelled dynamites are turned over at regular intervals to reverse the settling flow.

Nitroglycerine is used in the preparation of vasodilators. A very dilute solution is used in medicine as a neurotic.

Synonyms: Blasting Oil, Explosive Oil, Glyceroltrinitrate, Nitroglycerine, Nitrospan, Nitrostat, Trinitroglycerin, Soup.

Nitroglycerin Dynamite – See *Dynamite, Nitroglycerin.*

Nitroglycol – Nitroglycol [C_2H_4 $(NO_3)_2$] is a colorless and odorless liquid, sweetish in taste, only slightly more viscous than water, specific gravity 1.482 at 15°C. It freezes at about -22.3°. It is finding increasing application as a less sensitive explosive than nitroglycerine, both alone and in admixture with that substance. Nitroglycol is an ingredient of non-freezing dynamites. It has many of the advantages of nitroglycerin and is safer to manufacture and handle. Its principal disadvantage is its greater volatility.

Nitroglycol has slightly larger energy content than nitroglycerin.

It is too volatile for use in double-base smokeless powder, for its escape by evaporation affects the ballistic properties. But its greater volatility does not affect the usefulness in gelatin dynamite. Of course, it is unsuitable for use during the warm season of the year in ammonium nitrate explosives.

Synonyms: Ethylene Dinitrate, Ethylene Glycol Dinitrate, and Glycol Dinitrate.

Nitroguanidine – Nitrated Aminomethanamidine. This high explosive is about as powerful as TNT. Nitroguanidine is generally used mixed with colloided nitrocellulose, in which form it yields a propellant powder. The propellant powder gives no flash from the muzzle of the gun, thus serving as a great advantage to the military.

Because of its low temperature of explosion (about 2,098°C), nitroguanidine is used in triple-base propellants. The addition of nitroguanidine makes the triple-base propellant practically flashless and less erosive than nitrocellulose-nitroglycerine (double base propellant) of comparable force.

It is used as an additional base of propellant. Because of its low temperature of explosion (about 2,098°C), nitroguanidine is used in triple-base propellants. It is used as a ‘cool propellant’ because of its low temperature of explosion (about 2,098°C), which does not erode gun bores or produce as much luminous flash as single (nitrocellulose) propellants.

It has also been used mixed with ammonium nitrate and paraffin wax as trench mortar ammunition.

Synonyms: *Cool Explosive*, *Flashless Explosive*, and *Picrite*.

Nitrohydrene – Nitrohydrene is composed of nitroglycerin and nitrosucrose. It is a powerful explosive, approximately as powerful as nitroglycerin, and is used to stretch glycerin supplies.

It is made up by dissolving up to 25% of sucrose in glycerin and nitrating the resulting mixture to give explosive oil. This procedure saves considerable quantities of glycerin.

Nitromethane – Nitromethane is a liquid compound which is obtained by nitration of methane with nitric acid above 400°C (750°F) in the vapor phase. It is sparingly soluble in water. This compound is of industrial interest as a solvent rather than as an explosive. However, it can be used as a fuel in two-component (binary) explosives and as rocket fuel. Nitromethane with 5% ethylenediamine, known Picatinny Liquid Explosive (PLX), is used to clean up mine fields. It was used in the USA for underground model explosions. It was also employed in stimulation blasting in oil and gas wells.

Also see: *Tetranitromethane.*

Nitropropane – A liquid fuel that can be combined with pulverized ammonium nitrate prills to make a dense blasting mixture.

Nitrostarch – Nitrostarch, $[C_6H_7O_2(ONO_2)_3]n$, is not a definite compound, but a mixture of various nitric acid esters of starch of different degrees of nitration. Nitrostarch, with nitrogen contents varying from 12 to 13.3% is prepared by nitration of starch. The empirical formula of the structural unit of nitro starch is $C_6H_7N_3O_9$. It is a pale yellow powder. It is insoluble in water; and unlike starch it does not gelatinize when heated with water. It dissolves in acetone and alcohol-ether mixtures.

Nitrostarch is a high explosive. It is more sensitive to shock and friction than TNT, but is less powerful. Of course, it is less sensitive than dry Guncotton or Nitroglycerin. It has a velocity of detonation of approximately 6,100 meters (20,000 feet) per second. It is highly inflammable and can be readily ignited by a slight spark resulting from friction; it then burns with explosive violence.

Nitrostarch resembles nitrocellulose in several respects. It is similar to nitroglycerine in function. It is used as a substitute in case of TNT shortage. It is used as an ingredient of blasting explosives, for quarrying, and as a demolition explosive. 'Non-headache' industrial explosives are based on nitrostarch. Owing to its poor stability, hygroscopicity, and difficulty in preparation its use has become limited.

Synonyms: *Starch Nitrate*, and *Xyloidine.*

No. 8 Detonator – See *Detonator, No. 8*

No. 8 Test Cap – Institute of Makers of Explosives No. 8 Test Detonator.

Nobel, Alfred – Alfred Nobel, the greatest inventive genius in the explosive field, was born in 1833 in Stockholm, Sweden. At an early age, he was interested in explosives, and he learned the fundamentals of engineering from his father. Alfred Nobel was a competent chemist.

Nitroglycerin, a much more powerful explosive that was discovered in 1846 could not be handled with any degree of safety, as it was extremely volatile. At that time the only dependable explosive used in the mines was black powder. Alfred Nobel along with his inventor father built a small factory in 1862 to manufacture nitroglycerine, and undertook research in the hope of finding a safe way to handle it.

In 1863 he invented a *practical detonator.* This detonator marked the beginning of Nobel's reputation as an inventor.

In 1865 Nobel invented an improved detonator called a blasting cap to provide a safe and dependable means for detonating high explosives. The blasting cap consisted of a small metal cap (a copper capsule) containing a charge of mercury fulminate that can be exploded by either shock or moderate heat. The modern use of high explosives started with the invention of the blasting cap. The invention of the blasting cap is an important development in the history of explosives.

Nobel's task of making *nitroglycerine* more stable so that it could be used as a commercially and technically useful explosive ended successfully with his second important invention of *dynamite* in 1867, which was easy to use and safe to handle. Nobel got the credit not only for nitroglycerine but dynamite, too. Dynamite established his fame worldwide.

He also continued to experiment in search of better fillers and in 1875 he discovered a gel upon mixing nitrocellulose with nitroglycerin. The gel was developed to *blasting gelatin,* and *gelatin dynamite.* Later in 1888 Nobel introduced *ballistite*, the first nitroglycerin smokeless powder and a precursor of *cordite.*

Introducing Ammonium nitrate in the explosive field was also an invention of Alfred Nobel. By combining it with dynamite he laid the foundation for the so-called safety and permissible ammonia dynamites.

Nobel gradually built up his empire of explosive and chemical factories. He built factories and laboratories in 20 different countries. At the time of his death, the total number of his worldwide factories manufacturing explosives and ammunition rose to about 90. Nobel not only invented the powerful explosives but also established the world's most prestigious prizes for intellectual services rendered to humanity.

Alfred Nobel died on December 10, 1896.

No-Fire Current – No-fire current means the maximum current that can be continuously applied to a bridgewire circuit without igniting the prime material.

Non-cap Sensitive Explosive – Any product generally used in blasting that cannot be detonated by a #8 blasting cap. This subclass of high explosives, called blasting agents, as mixed and packaged for storage and transportation, cannot be detonated by a #8 blasting cap in a specific test. In normal commercial practice, a cap sensitive booster is used to detonate a non-cap sensitive explosive material.

Non-cap sensitive explosives are a class for storage and transportation. Unlike sensitive high explosives, these can be stored and transported under different regulations.

Non-freezing Dynamite – See *Dynamite, Non-freezing.*

Non-delay Fuse – Fuse that functions as a result of inertia of the firing pin (or primer) as the missile is retarded during penetration of target. The inertia causes the firing pin to strike the primer (or primer the firing pin) initiating fuse action. This type of fuse is inherently slower in action than the super quick or instantaneous fuse, since its action depends upon deceleration (retardation) of the missile during penetration of the target.

Non-electric Delay Blasting Cap – A blasting cap with an integral delay element in conjunction with and capable of being detonated by a detonation impulse or signal from miniaturized detonating cord. [Quoted from US Code.]

Non-Electric Blasting Cap – It is a blasting cap designed for and capable of detonation from the sparks or flame from a safety fuse inserted and crimped into the open end.

Non-Electrical Detonator – Non-electrical detonators or fuse caps are thin metal or paper cylindrical shells, open on one end for the insertion of safety fuse, which contain various types of primary and secondary explosives. They are sensitive to heat, shock and crushing, and are designed to be initiated with safety fuse or detonating cord. They are normally rated as #8 strength. All detonators of this type are instantaneous and therefore do not have a delay elements.

It is used to initiate other explosives, detonating cord and shock tube.

Non-electric Detonator – The detonator that does not necessitate the use of electric energy or safety fuses to function. It is used to initiate explosives and work with fast burning pyrotechnics.

Also see: S*hock Tube System.*

Non-nitroglycerin Dynamite – A composition explosive having similar characteristics to dynamite, but without nitroglycerin is erroneously called non-nitroglycerin dynamite. Theoretically, nitroglycerin is essentially a constituent of dynamite. But some manufacturers of dynamite have products in which nitroglycerin has been substituted by non-headache producing high explosives such as nitrostarch or other compounds. Non-nitroglycerin dynamite cartridge material is generally packaged on standard dynamite machinery.

Because of its high sensitivity to impact, nitroglycerin is usually avoided in military explosives. Of course, the non-nitroglycerin stick explosives for military purposes that perform like dynamite are less sensitive than nitroglycerin dynamite.

Non-Seated Explosion – The explosion in which usually there is no physical evidence of a single location where the explosion originated. Non-seated explosions are generally a result of the ignition of diffuse fuels such as gases (Industrial Gases LPG, Natural Gas, Sewer Gases), vapors of pooled liquids (gasoline vapors, lacquer thinner, MEK, etc), and dusts. It should be noted that the epicenter of the non-seated explosion is not the location of the fuel gas leak, but it is at the source of ignition.

Characteristics of the non-seated explosions are widespread fuels, deflagration reaction with subsonic speed, moderate rates of pressure rise, and no crater formation.

Even though there is no crater formation in the rarefied explosions, by careful analysis of the scene often there is a definite overpressure direction that can be detected. Also, the epicenter is where the source of ignition is and not the location of the fuel gas leak.

Antonym: *Seated Explosion.*

Normal Charge – Normal charge means propelling charge employing a standard amount of propellant to fire a gun under ordinary conditions, as compared with a reduced charge or a super charge used in special circumstances.

Novelty – Novelty denotes a device containing small amounts of pyrotechnic and/or explosive composition, but not falling under the category of consumer fireworks. Such devices produce limited visible or audible effects. Examples are poppers, snakes, snappers, and tanks.

NS Gelignite – A nitroglycerine gelatin explosive containing sodium nitrate as its main oxidizing ingredient.

Nuclear Bomb – See *Nuclear Weapon.*

Nuclear Explosion – The explosion that may occur as a result of the release of internal energy during an uncontrolled nuclear reaction, either fission or fusion or both. Nuclear explosions like thermonuclear reactions on the sun's surface have always been present in the universe.

Nuclear explosions produce both immediate and delayed destructive effects. The energy produced from a nuclear explosion is a million to a billion times greater than that of a chemical explosion. A nuclear explosion would be fatal to anybody near the explosion, because of production of the heavy flux of neutrons. Even those who are some distance from the explosion would be harmed by the gamma radiation. A nuclear explosion also discharges intense infra-red and ultra-violet radiation. Although the shockwaves produced in an atomic explosion are similar to those of a chemical explosion, it will last longer and have a higher pressure in the positive pulse and a lower pressure in the negative phase. Intense infra-red and ultra-violet radiation is released in an atomic explosion. The delayed effects are radioactive fallout and other possible environmental effects inflict damage over an extended period ranging from hours to centuries, and can cause adverse effects in locations very distant from the site of the detonation.

Synonym: Atomic Explosion.

Also see: *Nuclear Weapon.*

Nuclear Explosive – Nuclear Explosive means an assembly containing fissionable and or fusionable materials, and main charge high-explosive parts or propellants capable of producing a nuclear detonation, such as a nuclear weapon or test device.

Nuclear Explosive Operation – Nuclear Explosive Operation (NEO) means any activity involving a nuclear explosive, including activities in which main charge high-explosive parts and pit are collocated.

Nuclear Weapon – See *Weapon, Nuclear.*

Number 8 Detonator – See *Detonator No. 8.*

Also see: Detonator.

Occupied Area – Any work area that can be reasonably considered integral to an explosives operating area, to which personnel are assigned or in which work is performed, however intermittently. Examples of areas to be considered as occupied are assembly/disassembly cells or bays, explosives operating bays, radiography control and film processing rooms, offices, break areas and rest rooms.

Octahydro-1,3,5,7-Tetranitro-1,3,5,7-Tetrazocine –

Structural chemical name of *HMX*

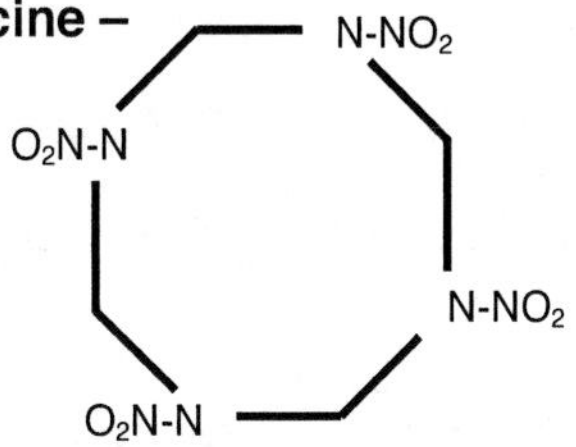

$N\text{-}NO_2$
$O_2N\text{-}N$
$N\text{-}NO_2$
$O_2N\text{-}N$

Octogen – A synonym of *Octahydro-1,3,5,7-Tetranitro-1,3,5,7-Tetrazocine.*

Also see: *HMX.*

Octol – Synonym of *Octolite.*

Octolite – An intimate mixture of *Octogen* and *TNT* in the ratios of 70/30 and 75/25. It is a secondary high explosive with a velocity of detonation 8,400 to 8,630 meters (27,500 to 28,300 feet) per second. It is excellent for blast effects. Octolite is used in projectile and bomb filler.

Synonym: Octol.

Oil, Blasting – See *Blasting Oil.*

-Oil, Di – See *Di-Oil.*

Oil, Explosive – Explosive Oil: 1. Liquid sensitizers for explosives, such as nitroglycerin, ethylene glycol dinitrate, and metriol trinitrate. 2. Synonym of Nitroglycerin.

Oil, Lamp, Safety – See *Safety Lamp Oil.*

-Oil, Mono – Mono-Oil: It is a mixture of about 62 percent ortho-, 34 percent para-, and 4 percent meta-nitrotoluenes. It is yellow oil with pungent odor. Its specific gravity is 1.16.

Mono-oil is used in low-freezing dynamites. It is also used as plasticizer for nitrocellulose. In blasting explosives it is used in combination with chlorates or nitrates. Such plastic explosives have the desirable property of completely filling the borehole by using only a light pressure. The oily nitro compound also serves as a protection against dampness, and it has the advantage of being an active ingredient in comparison with paraffins and greases which have previously been used as plasticizers.

Oil of Mirbane – Commercial name for *Nitrobenzene.*

Oil, Safety Lamp – See *Safety Lamp Oil.*

Oil Well Cartridge – Tubular explosive devices consisting of a shell of thin fiber, metal or other composition. It contains only propellant powder. Jet perforator and commercial shaped charge are not included in this definition.

Operational Grenade – A grenade that contains a bursting charge.

Operation, Contact – Contact Operation: An operation in which both the operator and the explosive item are present without any operational shield.

Operation, Remote – Remote Operation: It means an operation performed in a manner that will protect personnel in the event of an accidental explosion. The objective is to maintain distance, a shield, barricades, or a combination thereof.

Operation, Shot Firing – Shot Firing Operation: In the order of sequence, shot firing operation includes the activities, namely: (i) priming a cartridge, (ii) charging and stemming a shot hole, (iii) linking or connecting the detonator wires in a round of shots, (iv) coupling a shot firing circuit to a detonator circuit, circuit tester or exploder, (v) testing a shot firing circuit, and (vi) firing a shot or round of shots.

Operational Shield – The expression 'Operational Shield' means the barricade constructed to protect personnel, material, or equipment from the effects of a possible fire or explosion occurring at a particular operation.

Open Air Explosion – A *vapor cloud explosion* is sometimes termed as open-air explosion.

Synonyms: *Unconfined Vapor Cloud Explosion, Vapor Cloud Explosion,* and *UVCE.*

Ordnance – Generally speaking, ordnance means the larger-size military guns and mortars. Ordnance equals munitions. Hence ordnance includes military material, such as combat weapons of all kinds with ammunition and equipment for their use. It is mentionable that ammunition is a generic term for all kinds of missiles, explosives and pyrotechnic devices. The term 'ordnance' also includes non-offensive military items.

Ordnance Disposal, Explosive – Explosive Ordnance Disposal: The term is used to describe military operations in which hazardous devices are rendered safe.

Acronym: *EOD.*

Ordnance, Unexploded – Unexploded Ordnance: Explosive weapons (bombs, shells, grenades, etc.) which have been primed, fused, armed, or otherwise prepared for action, and which have been fired, dropped, launched, projected, or placed in such a manner as to constitute a hazard to operations, installations, personnel, or material and remains unexploded either by malfunction or design or for any other cause. Unexploded ordnances still pose a risk of detonation, decades after the battles in which they were used.

Acronym: UXO.

Ordinary Blasting Cap – A detonator initiated by a safety fuse.

Synonym: *Fuse Cap,* and *Fuse Detonator.*

Organic Nitrate – Organic nitrates are usually termed nitro compounds. However, in almost every case they are nitric acid esters of an organic material. These compounds are a combination of the nitro ($^{-}NO_2$) group and an organic radical. In general, these nitro compounds, or organic nitrates, are flammable and explosive. They can be detonated by means of boosters, by means of shock, temperature, friction or a combination of any of these methods. Some of them are more useful for one purpose than another.

Output Characteristics – The characteristics of an explosive component which determine the form and magnitude of the energy released when the component functions.

Overpressure – The pressure developed above atmospheric pressure at any stage or location is called the overpressure. Overpressure is sometimes used to describe exposure of equipment to pressures in excess of the design pressure.

Also see: *Peak Positive Overpressure, Side-on Overpressure, Reflected Overpressure, Blast wave, and Shock wave.*

Oxidant – Any gaseous material that can react with a gas, dust, or mist to produce combustion. Oxygen in air is the most common oxidant.

Oxidizer – An ingredient in an explosive or blasting agent, which supplies oxygen to combine with the fuel to form gaseous or solid products of detonation. Ammonium Nitrate is the most common oxidizer used in commercial explosives.

Oxidizing Agent – An element or a compound, which is capable of adding oxygen or removing hydrogen, or one that is capable of removing one or more electrons from an atom or group of atoms.

Oxidizing Material – Synonym of Oxidizer.

Oxygen Balance – The theoretical percentage of oxygen in an explosive material, or ingredient thereof, which is needed to produce products of full oxidation as reaction products. A mixture containing excess oxygen has a positive oxygen balance. One with excess fuel has a negative oxygen balance.

Oxygen Explosive, Liquid – Liquid Oxygen Explosives (LOX): A high explosive composed of absorbent carbonaceous or some other flammable material, such as cellulose, lamp-black, soot, cork-meal, or industrial earth, saturated with liquefied oxygen, and with or without various metallic powders.

It was used in coal mining, quarrying, strip mining, and open-cut ore mining. Its use underground or in confined places is not recommended due to the evolution of a large percentage of carbon monoxide.

It is stronger than black powder. It can be used for some purposes instead of gunpowder and dynamite. It can also be used in mine rescue apparatus and for air-ships. However, this explosive is rarely used today.

This explosive has some safety advantages. Since neither of the two components is an explosive, there is no risk before the two components are mixed together. Mixing is done at the last moment before firing. On the other hand there are some disadvantages with the made up explosive. It is very flammable, and when it catches fire it will usually detonate. This can be very dangerous to personnel and equipment. Liquid oxygen explosives are not stored as they deteriorate rapidly and lose a great deal of their explosive power in a short time.

The destruction of liquid oxygen explosives is not usually necessary, since they are made up as they are used. However, if enough time elapses after making up such an explosive, it will lose much of its explosive power, as the liquid oxygen evaporates, and then with proper precautions the remaining material can be disposed of.

Oxyhydrogen Flame – A flame obtained from the combustion of a mixture of oxygen and hydrogen.

P1 – The first and strongest class of *permissible/permitted explosives.*

P2 – *P1* explosives enclosed in a sheath to give increased safety. Originally called sheathed explosives.

P3 – A Class of *permissible/permitted explosives* for general use with instantaneous detonators.

P4 – A Class of *permissible/permitted explosives* particularly for ripping with short delay detonators.

P5 – A Class of *permissible/permitted* explosives particularly for firing with short delay detonators in solid coal.

Paper Match – A paper match that strikes only on a specially prepared surface to ignite.

Particulate Explosion – The explosion that can happen from the two-phase heterogeneous explosive mixtures of liquid or solid particles and air, gas or vapor, as and when they come in contact with the spark of any source of ignition. The particulate explosion is a type of *rarefied explosion.* Particulate explosions can again be sub-divided into *dust explosion* and *aerosol explosion.*

Also see: *Aerosol Explosion*, *Dust Explosion*, *Heterogeneous Explosion*, and *Rarefied Explosion.*

Paste – The initial mixture of guncotton and nitroglycerine, in the manufacture of double base propellant manufacture.

PBX – Acronym for *Plastic/Polymer Bonded Explosive.*

PE – Abbreviation of 'plastic explosive'.

PE 4 – The British equivalent of *C-4.* 88% RDX, 12% plasticizer (8% paraffin, 3% Lithium Steareate, 1% Pentaerythritol Dioleate).

Peak Pressure – Peak pressure means instantaneous maximum pressure developed in the chamber of a gun by burning propellant. Pressure immediately preceding an expanding shock wave is also termed as peak pressure.

PELAN – Acronym for Pulsed Elemental Analysis with Neutrons. It is a method for detection of explosives.

Pellet Powder – Pellet means a consolidated charge of explosive; and *black powder* pressed into cylindrical pellets is known as pellet powder.

Pentaerythroltetranitrate – A secondary high explosive compound, pentaerythrol tetranitrate represented by the empirical formula $(CH_2.O.NO_2)_4$. Its color is white unless dyed. A high explosive of exceptional brisance, Pentaerythroltetranitrate (PETN) is a well-known military explosive. In its raw crystalline state it is quite sensitive, particularly to blows, and the presence of grit increases impact sensitivity. Since PETN is more sensitive to shock or friction than TNT, it is primarily used in small caliber ammunition. It is one of the most powerful and violent high explosives in use, velocity of detonation being 8,300 meters (27,200 feet) per second. When used in a detonating cord, it has a detonation velocity of 6,400 meters (21,000 feet) per second and is relatively insensitive to friction and shock. It has 140% the power of TNT; and it is comparable in force to *nitroglycerine* and *RDX*.

$$\begin{array}{c} CH_2ONO_2 \\ | \\ O_2NOH_2C-C-CH_2ONO_2 \\ | \\ CH_2ONO_2 \end{array}$$

PETN

PETN is very stable, insoluble in water, sparingly soluble in alcohol, ether and benzene, and soluble in acetone and methyl acetate.

It is used in detonating cord, boosters, detonators, blasting caps and as a constituent of *pentolite*. It is also used as a base-charge in anti-aircraft shells and mixed with TNT (70-30) in mines, explosive bombs and torpedoes. It is a very effective demolition explosive. It is also used in blasting caps.

Acronym: PETN.

Synonyms: Penthrite, and Pentrit.

Penthrite – Synonym of *PETN.*

Pentolite – A detonating explosive that is an intimate mixture of equal proportions of *PETN* and *TNT*. It is light yellow in color. Often used in grenades. Its rate of detonation is 7470 meters (24,500 feet) per second. Pentolite may be melted and cast in the container. When cast, it is used as a cast primer. It is used as boosters, the main bursting charge in small shells and shaped charges. Pentolite is often used in grenades.

Pentolite Booster – A *booster* consisting of a mixture of *PETN* and *TNT*.

Pentrit – Synonym of *PETN*.

Perchlorate – Perchlorates are combinations of a metal and the monovalent $^{-}ClO_4$ radical. Perchlorates are powerful oxidizing agents, and when mixed with carbonaceous material, they form explosive mixtures. Additionally, perchlorates constitute a fire and explosion hazard when mixed with finely divided metals. This is also true for sulfur, powdered magnesium and aluminum.

Perchlorates in general are less sensitive than chlorates. Although pure perchlorate is not especially sensitive, the presence of impurities can render perchlorates very sensitive to shock and friction. Perchlorates are used in detonating and pyrotechnic compositions and in the manufacture of 'perchlorate explosives'. They are less sensitive to detonation and therefore safer than 'chlorate explosives', which they closely resemble.

Later forms of this explosive contained a further addition of small amounts of nitroglycerin or collodion cotton. A typical formulation used in ore mining and stone quarries is:

Potassium perchlorate	65%	Nitroglycerin	5%
Aromatic compounds	25%	Organic material	5%

For environmental reasons, there is very limited commercial use of perchlorate explosives today.

Percussion – A method of initiating an explosive item by a sudden sharp blow. The striking of a gun hammer on fulminating powder is such a blow.

Percussion Bullet – A bullet containing a substance that is exploded by percussion.

Synonym: Explosive Bullet.

Percussion Cap – A device used in firearms. It serves as an igniting element in small arms cartridges.

The percussion cap consists of a small disposable metal cap or cup (usually of copper), containing an *initiating* explosive, such as *fulminate of mercury*, which explodes upon being struck by the blow of the firing pin against the *primer cap*, thus initiating the explosion of the main charge to propel the bullet. That means the percussion cap is a firearm primer.

The percussion cap may also be defined as a detonator that explodes by impact.

Also see: *Percussion Cap System.*

Synonyms: Cap Type Primer, Percussion Primer, and Small Arm Primer.

Percussion Cap System – The Reverend John Forsyth developed the percussion cap system in 1850. It has a cylindrical cap or cup, which leads straight into the gun barrel. It contains a primary explosive, generally fulminate of mercury, which causes a spark when the hammer strikes the cap. This spark ignites the main black powder charge in the bore to propel the bullet. This system allowed for quicker reloading and worked better in poor weather conditions compared to the flintlock system. Before the crucial invention of percussion caps, firearms used igniters with flints or matches to set fire to a pan of gunpowder. The percussion cap firing mechanism has given a weapon precision and reliability because of the certainty of the percussion cap to explode when struck. The percussion cap was the key to making reliable rotating-block guns (revolvers), which fire reliably.

Percussion Composition – Percussion composition means the high-explosive powder, which is ignited in some types of firearms by the blow of the firing pin against the primer cap.

Percussion Fuse – A fuse in which the ignition is produced by a blow on some fulminating compound.

Synonym: Impact Fuse.

Percussion Lock – Percussion lock means the lock of a gun that is fired by percussion upon fulminating powder.

Percussion Match – A match ignited by percussion.

Percussion Powder – Powder so composed as to ignite by slight percussion.

Synonym: *Fulminating Powder.*

Percussion Primer – Cylindrical cap or cup containing a small charge of initiating explosive that may be set off by a blow. A percussion primer is used in all fixed and semi-fixed ammunition and in certain types of separate-loading ammunition to ignite the main propelling charge.

Percussion Wave – A *shock wave* from an explosion or a blow.

Perforated Powder – In a propellant, if the grains are made so that each grain has a hole in it, it can then be a neutral powder because, as it burns the surface area remains about the same. These are called perforated powders. Propellants can be made to perform differently by varying the physical characteristics of the individual grains because burning is mostly a surface phenomenon.

Also see: *Multi-Perforated Powder.*

-perforated, Multi, Powder – Multi-perforated Powder: See *Powder, Multi-perforated.*

Permanent Blasting Wire – It is a permanently mounted insulated wire used between the electric power source and the electric blasting cap circuit.

Permanent Magazine – See *Magazine, permanent.*

Permanganate – Permanganates contain the nono-valent $^{-}MnO_4$ radical. These are all powerful oxidizing agents. Such materials as silver permanganate and other metallic permanganates may detonate when exposed to high temperatures, or when they are involved in fires, or severely shocked.

Also see: *Potassium Permanganate*.

Permissible Diameter (Smallest) – The smallest diameter of a permissible explosive, as approved by the US Mine Safety and Health Administration (MSHA).

Permissible Dynamite – A dynamite composition in which the ammonium nitrate is the major constituent, and such composition also contains a flame coolant to lower the temperature of the explosion, reduce the amount of toxic fumes evolved, and reduce the flame. Certain amounts of salts with water of crystallization or salts that give off a large amount of gas on explosion, such as bicarbonates and oxalates, are added for the purpose of cooling the explosion gases.

Permissible Ammonia Dynamite – See *Dynamite, Permissible Ammonia.*

Permissible Explosive – An American term for an explosive with a low flame output that is specifically authorized for use in gassy underground coalmines. Permissible explosives produce little or no flame and explode at low temperatures to prevent secondary explosions of mine gases and dust. One such important explosive used in mining, is called ANFO.

Permissible Gelatin Dynamite – See Dynamite, Permissible Gelatin

Permissible Individual Maximum Pressure – For any type gun, that value which should not be exceeded by the maximum pressure developed by an individual round under any service condition.

Permitted Explosive – A British term for an explosive with a low flame output that is specifically authorized for use in gassy underground coalmines. The permitted Explosives produce little or no flame and explode at low temperatures to prevent secondary explosions of mine gases and dust. One such important explosive is used in mining, called ANFO.

Peroxide – Peroxide is a compound that contains the oxygen-oxygen linkage as follows:

$$\begin{array}{l} —O \\ \;\;\;| \\ —O \end{array}$$

The organic peroxides are a sub-class of oxidizing substances, being the class 5 dangerous substances.

The peroxides are all-powerful oxidizing agents. However, concentrated hydrogen peroxide is not only a powerful oxidizing agent, but it can decompose so rapidly as to detonate. This explosion is of the high explosive type, very powerful and very fast. Hydrogen peroxide in concentrated form has found use in rocket fuel manufacture, where advantage is taken of its high energy of decomposition. Concentrated hydrogen peroxide, from 30-100%, must be kept free from silica, manganese, and organic materials as well as many other materials in order for it to remain stable. It should be kept cool and should not be subjected to excessive shocks or vibrations.

Personnel Barrier – A device designed to limit or prevent personnel access to a building or an area during hazardous operations.

PETN – Acronym for *Pentaerythrol tetranitrate.*

Petrol Bomb – Synonym of *Molotov Cocktail.*

Petryl – Petryl is the synonym for the chemical compound *Trinitrophenylnitramine Ethyl Nitrate.*

Phlegmatise – To add a proportion of oil or other ingredient to render an explosive insensitive.

Phlegmatiser – A solid or liquid, non-explosive or explosive substance which is added to explosive substances with a view to reducing their sensitivity to heat and impact, thereby assisting in ensuring safety during their transport.

Phosphorescence – Moist yellow phosphorus emits a blush glow of light without evolution of much heat. This is a case of combustion without heat.

Phosphorous, White – See *White Phosphorous.*

Photochemical Explosion – Some chemical reactions are initiated, assisted, or accelerated by exposure to light. Hydrogen and chlorine combine very slowly in dark but they combine explosively on exposure to sunlight. Such act of explosively reacting or combining on exposure to light is termed photochemical explosion

Photo-flash Bomb – An explosive article dropped from aircraft with a view to providing brief, intense illumination for the purpose of photography.

Photoflash bombs are loaded with flashlight powder, which is similar to black powder as to hazards in handling and storage.

Photo-flash Powder in units – These consist of a pyrotechnic charge which, when ignited, produces a light of sufficient intensity and duration for photographic purposes, or for stage effects.

Physical Explosion – A physical explosion is an explosion in which chemical reaction plays no significant part. Examples of physical explosions are disintegration of vessels containing compressed gases or liquefied gases (strictly speaking liquefied vapors), bursting of boilers, massive electrical discharges, reactions between molten metals and water, reactions between cryogenic liquids and water, volcanic explosion, or a meteor hitting Earth. Some physical explosions, such as electrical discharges, a meteor hitting Earth, and volcanic explosion have always been present in the universe.

There are a number of types of physical explosion. Rapid phase transitions, which occur when a liquid comes into contact with a much hotter liquid or solid, is a category of physical explosion. The second one is that in which there is loss of containment of a compressed gas. The third arises from loss of containment of a liquefied vapor.

Synonym: Mechanical Explosion.

Picatinny Liquid Explosive – Picatinny liquid explosive is a mixture of nitromethane with 5% ethylenediamine and is used to clean up mine fields.

Picric Acid – True chemical name of picric acid is 2, 4, 6-trinitrophenol. It is a secondary high explosive represented by the empirical formula $C_6H_3N_3O_7$. Bright yellow crystals of picric acid are obtained by the nitration of carbolic acid (phenol). Picric Acid is one of the oldest explosives. It is more powerful than trinitrotoluene. In the cast form it is known as lyddite.

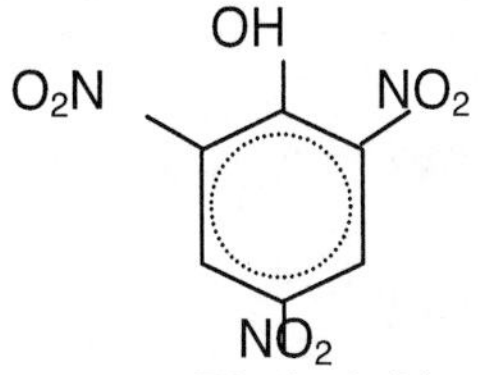

OH
O_2N NO_2
NO_2

Picric Acid

A number of salts of picric acid, known as picrates, are more sensitive explosives. This is particularly true of copper, lead and zinc. That is why picric acid has to be kept out of contact with metals. Its ability to form the sensitive picrates readily in contact with metal has somewhat limited its usefulness as an explosive. The use of picric acid was largely superseded by TNT, and by TNT mixed with ammonium nitrate.

It was used in early artillery shells as the bursting charge, and it is also useful as a primer and initiator. It has been used extensively in the form of mixtures with other nitro compounds. The mixtures, which have a lower melting point than picric acid, cause the mixture to be able to be melted and cast at temperatures below 100°C, the melting point of picric acid being 122°C. The mixtures are usually more practical for use because of the hazard involved in melting straight picric acid. Guanidine picrate is an example of a picrate used in shells instead of pure picric acid.

Picric acid is used in dying. It is a well-known direct yellow dye for silk, and stains parts of the body with which it comes in contact.

Picric acid has also found use as a booster explosive and as a substitute for a part of the mercury fulminate charge in detonators. It is not widely used today, because it is fairly sensitive and better substances are available for most applications. Picric acid is dangerous when it deteriorates.

Picrite – Synonym of *Nitroguanidine.*

Picryl Sulfide – See *Hexanitrodiphenyl Sulfide.*

Pillar and Stall Mining – A method of mining in which pillars are left to support the roof.

Pin Puller – A mechanical device in which a pressure cartridge causes a pin or piston to retract, usually against a side load.

Pin Pusher – A simple improvised explosive device in pressure cartridge drives a pin or piston along its central axis. In this device a piece of pipe is filled with an explosive material.

Pinwheel – See *Catherine Wheel, the synonym of Pinwheel.*

Pipe Bomb – A simple type of improvised explosive device in which a piece of pipe is filled with an explosive material. The pressure of the burning explosive material ruptures the pipe resulting in a sudden explosion. The rupturing pipe creates fragmentation leading to injury.

Pistol – A small arm whose chamber is integral with the barrel. This is opposed to a revolver. A pistol is held and fired with one hand.

Pistol, Very – See *Very Pistol.*

Pit Magazine – An improvised magazine for storing explosives temporarily. This magazine consists of an underground damp-proof pit/cavity, the floor of which is either cement or covered wooden planks. Essentially the explosives in the pit must be covered with tarpaulin awning so as to be protected against sun and rain, and the margin of the pit is so raised as not to allow rainwater to drain inside it. The storage should be fenced off all around to prevent unauthorized persons from entering the premises.

Plastic Bonded Explosive – Synonym of *Plastic Explosive.*

Plastic Explosive – An explosive material in flexible or elastic sheet form, formulated with one or more high explosives which in their pure form have vapor pressure less than 10-4 Pa at a temperature of 25°C, is formulated with a binder material, and is as a mixture malleable or flexible at normal room temperature. They consist of high brisance explosives such as RDX or PETN combined with plasticizers. From the above definition, the plastic explosive has the following criteria:

a. It consists of one or more high explosives;

b. The high explosive(s) have, in the pure form, a vapor pressure less than 10-4 Pa at a temperature of 25°C;

c. The high explosive(s) are formulated with a binder material; and

d. as a mixture, it is malleable or flexible at normal room temperature.

Plastic explosives, capable of being molded into desired shapes, have excellent chemical stability, good explosive properties, high thermal output sensitivity, high mechanical strength, and relative insensitivity to handling and shock.

The materials contain a high percentage of basic explosives of high brisance, e.g. RDX, HMX, HNS, or PETN, in a mixture with a polymeric binder (plasticizer). The mixture is then cured into the desired shape. Other components such as aluminum powder can also be incorporated in this explosive. The products obtained can be of any desired size, and specified mechanical properties can be imparted to them, including rubber-like elasticity. They can also be shaped into foils. Of particular importance for tactical operations are the 'sheet explosives', which are made with PETN or RDX, depending on the product.

These explosives are easy to use by non-experts. Propellant charges for rockets and guns have also been developed by compounding solid explosives such as nitramines (e.g. Cyclonite) with plastics. Plastic explosives and plastic propellants are of interest, if low thermal and impact sensitivity is needed.

Synonym – See *Composition C4.*

Platonization – The addition of ingredients, in propellants, to produce a low-pressure index over a working range of pressures.

Plume – Plume means vapor cloud of a hazardous chemical.

Plunging Fire – Gunfire that strikes the earth's surface at a high angle.

PLX – Acronym for Picatinny Liquid Explosive.

Pneumatic Placing – Pneumatic placing means the loading of blasting agent or explosive into a borehole using compressed air as the loading force.

Point Detonating Fuse – Fuse, located in the nose of a projectile, which is initiated upon impact.

Point, Control – See *Control Point.*

Point, Fire – See *Fire Point.*

Point, Firing – See *Firing Point.*

Point, Flash – See *Flash Point.*

Polymer Bonded Explosive – See *Plastic Explosive.*

Pool Fire – The combustion of material evaporating from a layer of liquid at the base of the fire.

Portable Magazine – A magazine that is not permanently fastened to a foundation. It is usually a substantial wooden box covered with galvanized sheet metal not less than twenty-six gauge, with the word "EXPLOSIVES" in letters not less than 15 centimeter (6 inches) in height painted conspicuously on the top. It is so constructed or secured as to make sure that it cannot be carried, lifted, or removed easily by unauthorized persons. Its capacity is usually limited to quantity of explosives required for safe and efficient operation.

Potassium Chlorate – Potassium chlorate, $KClO_3$ is the principal component of Chlorate Explosives. It is an important component of primer formulations and pyrotechnical compositions, in particular match heads.

Potassium chlorate is readily soluble in hot water, sparingly soluble in cold water, and insoluble in alcohol.

Potassium Nitrate – Potassium nitrate, KNO_3 is used as a component in black powder, pyrotechnical compositions, and in industrial explosives. It can also be used in tobacco treatment.

Potassium nitrate is readily soluble in water, sparingly soluble in alcohol, and insoluble in ether.

Synonym: Saltpeter.

Potassium Permanganate – Potassium Permanganate ($KMnO_4$) is dark purple crystals with a blue metallic sheen. It is a powerful oxidizing agent. It possesses a moderate fire hazard by chemical reaction with deducing agents. Potassium permanganate is spontaneously flammable on contact with glycerin and ethylene glycol. It also possesses a moderate explosion hazard when shocked or exposed to heat.

Potassium Perchorate – Potassium perchlorate is prepared by reacting a soluble potassium salt to sodium perchlorate or perchloric acid. Potassium perchlorate is insoluble in alcohol but soluble in water. It is used the manufacture of pyrotechnics.

Powder – The meaning of the term 'powder' is wide and varied. The term is used to mean:

a. A synonym designating any explosive, irrespective of type.
b. A generic name for propellant explosives.
c. The term is synonym of black powder.
d. Any solid explosive.
e. An explosive (or propellant) in the form of powder or small granules.

f. A common synonym for explosive materials.
g. A mixture of potassium nitrate, charcoal, and sulfur in a 75:15:10 ratio, which is used in gunnery, time fuses, and fireworks.
h. An explosive mixture used in gunnery, blasting, etc.

Powder, Ajax – See *Ajax Powder.*

Powder, Aluminum – See *Aluminum Powder.*

Powder, Amid – Amid Powder: Synonym of *Amidpulver.*

Powder, Ball – See *Ball Powder.*

Powder, B-black – See *B-black powder.*

Powder, B-blasting – See *B-blasting powder*.

Powder, Black – See *Black Powder.*

Powder, Black Blasting – See *Black Blasting Powder.*

Synonym: *Blasting Powder.*

Powder, Blank-fire – Blank-fire Powder: Synonym of *EC Smokeless Powder.*

Powder, Blasting – See *Blasting Powder.*

Synonym: Black Blasting Powder.

Powder, Boom – Boom Powder: A pyrotechnic ignition mixture aimed at producing incandescent particles. A typical boom composition is: Iron Oxide 50 parts, Titanium 32 parts, Zirconium 17 parts and cellulose nitrate (as a binder) 1 part by weight.

Powder Cake – Nitrocellulose impregnated with not more than 60% of nitroglycerin or other liquid organic nitrates or mixtures of these.

Powder Chest – A substantial, nonconductive portable container equipped with a lid and used at blasting sites for temporary storage of explosives.

Powder, EC – See *EC Powder.*

Powder, EC Smokeless – See *EC Smokeless Powder.*

Synonyms: Blank-fire Powder, and E.C. Blank Fire.

Powder Factor – The amount of explosive used per unit of rock. In other words, it is the ratio of the weight of the explosive to that of the rock burden, usually expressed as pounds per ton or as pounds per cubic yards. Powder factor is an important parameter in blasting.

Synonym: Explosive Loading Factor.

Powder, Flash – See *Flash Powder.*

Powder, Fulminating – See *Fulminating Powder.*

Powder, Fuse – Fuse Powder: A very fine black powder used in safety fuse. The fineness is 140-mesh or finer.

Powder, Giant – Giant Powder: Synonym of *Dynamite.*

Powder, Giant Low-Flame – See *Giant Low-Flame Powder.*

Also see: *Hydrated Explosive.*

-powder, Gun – See *Gunpowder.*

-powder, Gun, Smokeless – See *Smokeless Gunpowder.*

Powder, L.D. – See *L.D. Powder.*

Powder, Nitrocellulose – See *Nitrocellulose Powder.*

Powderman – A person who may charge or load blastholes with explosive charges but may not fire the blast.

Compare: *Shot Firer.*

Powder Monkey – Synonym of *Blaster.*

Powder, Multi-perforated – Multi-perforated Powder: In a propellant powder if the grains of the powder are made so that each grain has multiple holes in it, it can burn progressively, meaning that they burn faster as they are consumed. This allows powders to develop more gases after a projectile has begun to move down the barrel of a gun. The powder is then known as multi-perforated powder. It is worth mentioning that propellants can be made to perform differently by varying the physical characteristics of the individual grains.

Also see: *Perforated Powder.*

Powder, Pellet – See *Pellet Powder.*

Powder, Percussion – See *Percussion powder.*

Powder, Perforated – See Perforated Powder.

Also see: *Multi-Perforated Powder.*

Powder, Propellant – Propellant Powder: Synonym of *Smokeless Powder.*

Powder, Pyro – See *Pyro Powder.*

Powder, Smokeless – Smokeless powder: Any propellant explosive based on nitrocellulose. This includes propellants with a single base (such as nitrocellulose powder), those with a double base (ballistite, cordite) and triple base propellant (NC/NG/nitroguaniding). Smokeless powder is the commonly used propellant.

Powder, Soda – See *Soda Powder.*

Powder, Sporting – See *Sporting Powder.*

Powder Train – 1. Train, usually of compressed black powder, used to obtain time action in older fuse types. 2. Train of explosives laid out for destruction by burning.

Power, Candle – Candlepower: It is the luminous intensity of pyrotechnic compositions.

Power Device Cartridge – An explosive device designed and used to accomplish mechanical actions other than that of propelling rockets or projectiles. It consists of housing with a charge of deflagrating explosive and a means of ignition. The gaseous products of the deflagration produce linear or rotary motion or function diaphragms, valves or switches.

Devices designed to start mechanical apparatus by means of propellant explosives. They consist of housing with a charge of deflagrating explosive and a squib or electric igniter.

Power Devices, Explosive – See *Explosive Power Device.*

-power, Fire – See *Firepower.*

Practical Detonator – In the course of the search for a safe way to control explosive detonation, *Alfred Nobel* in 1863 invented a detonator consisting of a wooden plug inserted into a larger charge of nitroglycerin held in a metal container. The explosion of the plug's small charge of black powder serves to detonate the much more powerful charge of liquid nitroglycerin. Alfred Nobel's first important invention is known as 'practical detonator' that marked the beginning of Nobel's reputation as an inventor as well as the fortune he acquired as a maker of explosives. In 1865 Nobel invented an improved detonator called a blasting cap.

Practice Ammunition – The ammunition that does not carry the main bursting charge, but normally includes a propelling charge, explosive train and a burster, termed as practice ammunition.

Practice Grenade – *See Grenade, Practice.*

Premature Firing – The detonation of an explosive charge before the intended time.

Presplitting – A technique of blasting that gives accurate finished contours.

Press Cake – A cake taken from the presses, in the manufacture of black powder.

Pressed, Dead – See *Dead Pressed.*

Pressing, Dead – See *Dead Pressing.*

Pressure Bomb – See *Ballistic Bomb*, the synonym of pressure bomb.

Pressure Burst – The rupture of a system under pressure, resulting in the formation of a blast wave and missiles, which may have the potential to cause damage.

Pressure Cartridge – An explosive item designed to produce momentary gaseous products of combustion under pressure for performing a mechanical operation.

Pressure Index – In propellants, the variation of burning speed following changes in pressure.

Pressure Pot – A system for loading ANFO into blast holes, in which ANFO is contained in a sealed vessel, to which air pressure is applied, forcing the ANFO through a semi-conductive hose and into the small diameter blast hole.

Synonym: *Pressure Vessel.*

Pressure Tank – Pressure tank means any large artificial receptacle in which a gas or liquid is held, transported or stored under a pressure greater than the atmospheric pressure.

Pressure Vessel – 1. Any closed metal container of whatever shape in which compressed gas is held, transported or stored. In other words, it is a metal container that is subjected to internal pressure. 2. Synonym of *Pressure Pot.*

Prill – An porous spherical form of a product, particularly ammonium nitrate.

Prilled Ammonium Nitrate – See *Ammonium Nitrate, Prilled.*

Primacord – Flexible fabric tube containing a filler of high explosive that is used to transmit a detonation from a detonator to a booster or bursting charge. Primacord is the trade name for one type of detonating fuse currently in use.

Also see: *Detonating Cord.*

Primadet – *Detonating cords* containing about 2 g of the high explosive *PETN* per meter inside a plastic-impregnated network are manufactured as 'Primadet'.

Also see: *Detonating Cord.*

Primary Blast – A term used in commercial blasting to describe a blast by which the original rock formation is dislodged from its natural location to facilitate subsequent handling and crushing.

Primary Explosive – Primary explosive is one which is highly sensitive to spark, flame, impact or friction, used in detonators and primers to initiate another less sensitive high explosive, namely the secondary explosive. In other words, the sensitive explosives used as initiating explosives are the primary explosives. Essentially the primary explosive when used in varying amounts in the form of detonator (blasting cap) and primers, it initiates the main charge of secondary high explosive. Common primary explosives are: Mercury fulminate, almost any azide including Lead azide, Lead picrate, Lead styphnate, DDNP, HMTD, and TACC.

In broad classification, a primary explosive falls under the high explosive (detonating explosive). It is a sub-class of detonating explosive.

Primary explosive is an alternative name for an initiating explosive. It is one of the first elements in an explosive train. A primary explosive does not burn but detonates if ignited. The reaction of a primary explosive starts with a deflagration, but within a few microseconds or less becomes a detonation.

Usually a primary explosive is not found in the raw form. It may be handled only if it is already pressed into capsules. The cylindrical capsules are generally made of aluminum or copper. Since it is reactive to different materials, care should be taken when handling a primary explosive when in the natural state. A primary explosive is highly brisant and must have a high triggering velocity.

Primary explosives are also filled in percussion caps mixed with friction agents and other components. Primary explosives are used as an additive in primers. Primary explosives are very expensive and are not used for rock-breaking on their own.

Synonyms: Initiating Explosive, and Primary High Explosive.

Primary Fragmentation – The act of producing fragments directly from the contents or casing of an explosive device.

Primary High Explosive – Primary High Explosive is usually shortened to *Primary Explosive.*

Primed Cartridge – Primed cartridge means a cartridge of explosives into which a detonator has been inserted.

Primed Empty Cartridge Shell – The expression 'primed empty cartridge shell' means empty cartridge shell containing an ignition primer.

Primed Empty Grenade – Primed empty grenade means empty grenade containing an ignition primer.

Primer – Literally means 'that which primes'. An instrument or device for priming is a primer. It is a device used to initiate the burning of a propellant charge by means of a flame. Its explosive train normally consists of a small quantity of initiating explosive, which when detonated ignites a small black powder booster which, in turn, ignites the black powder igniter. In short, it is a firing device and an explosive charge used to set off the main charge.

It may also be termed as any igniter by which an explosive charge is ignited. The term primer describes a device that contains explosives, not the explosives themselves. A cartridge or container of explosives into which a detonator or detonating cord is inserted or attached is an example of a primer.

Relatively simple and sensitive components are used as the prime elements in more complex devices, constituting the starting point of an explosive train or a train of ignition. They may be actuated by friction, percussion pressure or electricity. In the case of cartridges for small arms, such elements constitute the complete means of ignition.

Primers can either detonate or deflagrate.

Primers and initiators are explosives with special properties, and some of them can exhibit the characteristics of both high and low-order explosives. Most of them are high explosives and they are much more sensitive to friction, heat and shock than other forms of explosives. In some of these materials, physical impact or vibration can cause an explosion. Some are strictly deflagrating, but when confined they detonate. They are usually used in small quantities to initiate an explosion in another explosive material. Some high explosives cannot be detonated without the use of some other explosive to set up an initial shock wave, and a few must have a booster charge.

In relation to firearms, it may be described as a small metal cup that contains a tiny explosive charge that is sensitive to impact. A primer is placed in the base of a shell casing to ignite the powder of the completed cartridge. The striking of a firing pin in the firearm detonates it.

All primers function in a similar manner when initiated. Primers are classified according to the method of initiation, such as electric primer, or percussion primer.

When a booster is armed with a detonator, it becomes a primer. When a primer has the correct dimensions and properties, it can function as a booster.

Primer, Capped – Capped Primer: A package or cartridge of cap-sensitive explosive, which is specifically designed to transmit detonation to other explosives and which contains a detonator. [Quoted from MSHA Regulations.]

Primer, Cap Type – Cap Type Primer: Cap type primer is a metal or plastic cap that contains a small amount of initiating explosive, which is readily ignited by shock. It serves as an igniting element in small arms cartridges. The purpose of cap type primer is to kindle small arms cartridges.

Primer, Cast – Cast primer means a cast unit of explosive used to initiate detonation in a blasting agent. Ideal example of cast primers is pentolite and composition B.

Primer Cord – A synonym of *Detonating Cord.*

Primer, Detonating – Detonating primer: A name applied for transportation purposes to a device consisting of a detonator and an additional charge of explosives, assembled as a unit.

Primer-Detonator – Assembly consisting of a primer and a detonator. It may also include a delay element.

Primer, Electric – Electric Primer: Metallic device containing a small amount of sensitive explosive or charge of black powder, which is actuated by energizing an electric circuit. It is used for setting off explosive or propelling charges.

Primer for Propelling Charge – It is an assembly combining in one metallic tube a primer element for ignition and an auxiliary charge of igniferous composition, used to ignite propelling charges for cannons, etc.

Synonym: *Tubular Primer.*

Primer, Igniting – See *Igniting Primer.*

Primer Mixture – An explosive mixture containing a sensitive explosive, usually the first element in an explosive train.

Primer, Percussion – See *Percussion Primer.*

Primer Seat – Primer location within the breech chamber of a gun, which uses separate loading ammunition.

Primer, Small-Arms Ammunition – See *Small-Arms Ammunition Primer.*

Primer, Tubular – Tubular Primer: Synonym of *Primer for Propelling Charge.*

Priming – The act of making something ready.

Priming, Axial – See *Axial Priming.*

Priming, Boiler – See *Boiler Priming.*

Priming Charge – In a detonator, the small amount of initiating explosive (primary explosive), which fires the small packet of secondary high explosive (the base charge). The priming charge may be fired electrically or non-electrically, which is the basis of two basic types of detonator structure.

Priming Tube – A small pipe, filled with a combustible composition for firing cannon.

Priming Wire – See *Wire, Priming.*

Product Lot Sampling – Tests conducted on a sample of a production lot to determine that the lot meets the specified dimensional and firing characteristics.

Progressive Burning – The term 'Progressive burning' is used to describe the burning of a propellant grain in which the reacting surface area increases during the combustion.

Progressive Granulation – Propellant grain that burns with a continually increasing surface until the grain is completely consumed.

Projectile – An object, such as a bullet or shell, which is propelled from a weapon by an explosive propelling charge. A bullet fired from a gun is a projectile.

Projectile, Dummy – Dummy Projectile: Shell having no explosive charge. It is used for practice and training purposes.

Projectile, Explosive – Explosive Projectile: It means a metal shell loaded with explosives for use in cannons.

Projectile, Smoke – See *Smoke Projectile.*

Proof Ammunition – Ammunition incorporating solid, blunt-nosed, steel or cast iron shot of inexpensive manufacture; used in proof firing of guns; used to simulate the weight of projectile designed for the gun in adjusting the charge weight or propellant.

Propagation Blasting – Explosive charges may detonate by an impulse received from adjacent or nearby explosive charges. The theory is applied in blasting. In propagation blasting, sensitive charges are spaced closely. The shock from the first charge propagates through the ground, setting off the adjacent charge, and so on. In this process only one detonator is required. Propagation blasting is primarily used for ditching damp ground.

Also see: *Sympathetic Detonation.*

Propagation of Detonation – Synonym of *Sympathetic Detonation.*

Propagation of Explosion – Synonym of *Sympathetic Detonation.*

Propellant – Propellant means explosives that propel, such as explosives that are used for propelling projectiles from guns; for propelling rockets and missiles, launching torpedoes, and launching depth charges from projectors; and for generation of gases for powering auxiliary devices are called propellants. Propellants have found their use in filling cartridges, shell cases, and solid fuel rockets. The term is also used to include the fuel and oxidant of rockets when these are separate.

A propellant is a low explosive; the term is very often used as a synonym of low explosive. A propellant may also be defined as an explosive that usually deflagrates (burns) and does not detonate. As a class it is one of the two classes of explosives, the other being detonating explosive (high explosive).

The prime objective of a propellant is to deflagrate. Here lies the difference between a high explosive and a propellant. Its low rate of combustion permits its use as a propelling charge. A propellant develops relatively high pressures without the presence of a higher velocity shock wave. This allows work to be performed by the pressure increase and does not cause fracturing of the containment chamber. For example, if RDX were used in a gun rifle, the barrel would shatter, whereas black powder burns such that the pressure buildup accelerates the propulsion of bullet out of the barrel. Most modern military low-order explosives are used as propellants.

Examples of propellant are rifle and gun ammunition. Commonly used propellants are: black powder, smokeless powder, nitrocellulose/nitroglycerin compositions.

Propellant Actuated Device – Propellant actuated device (PAD) is a mechanical device actuated by a contained or inserted propellant charge.

Propellant Composite – Propellant compositions commonly contain additives, which affect the performance of the propellant.

Propellant, Composite – See *Composite Propellant.*

Propellant, Double Base – Double base propellant: One of the three classes of propellants, double base propellants are nitrocellulose powders to which nitroglycerin has been added to make them more powerful. Double base propellants are based on the active ingredients of nitrocellulose and nitroglycerine.

Propellant Explosive – An explosive that is used for propulsion purposes and which normally functions by deflagration. As a class, it is one of the two classes of explosives, the other being detonating explosive (high explosive). A propellant explosive is a low explosive. The term is very often used as synonym of low explosive.

Propellant, Rocket – See *Rocket Propellant.*

Propellant Powder – Synonym of *Smokeless Powder.*

Propellant, Single Base – See *Single Base Propellant.*

Propellant, Smokeless – See *Smokeless Propellant.*

Propellant, Solid – Solid Propellant: Specifically, a rocket propellant in solid form, usually containing both fuel and oxidizer combined or mixed and formed into a monolithic (not powdered or granulated) grain.

Propellant, Triple-Base – See *Triple-Base Propellant.*

Propelling Charge – The explosive charge that is burned in a weapon to propel a projectile. Burning of the confined propelling charge produces gases whose pressure forces the projectile out. The propelling charges may be in any physical form for cannon or small arms, for explosive power devices or for rockets, irrespective of whether they are military or commercial. In other words, cast or pressed charges of propellant explosive are called propelling charge. Pyrotechnic devices are not included in this definition.

Also see: *Propellant.*

Proximity Fuse – A fuse designed and used to detonate a bomb, charge, mine, or projectile as and when activated by an external influence in the close vicinity of the target.
Purging – Purging is the process of supplying an enclosure with a protective gas at a sufficient flow and positive pressure to reduce the concentration of any flammable gas or vapor initially present to an acceptable level.

Pyro – Synonym of *Pyrocotton.*

Pyrocellulose – Synonym of *Pyrocotton.*

Pyrochemical – The term 'pyrochemical' may be used in lieu of pyrotechnical, especially in regard to the chemical reactions of pyrotechnics.

Pyrocore – A flexible explosive cord similar to *MDF* except that the high explosive core is modified to promote ignition at the speed of detonation. It is a high velocity ignition propagation fuse (detonating).

Pyrocotton – Pyrocotton is a form of nitro-cellulose, having essentially the composition of Decanitrocellulose $C_{24}H_{30}O_{10}(NO_3)_{10}$. This material contains about 12.6% nitrogen. This form of nitrocellulose contains less nitrogen than guncotton. It is white colored pulped fibers. Pyrocotton is soluble in ether-alcohol (2:1) and in acetone. When used as a propellant in cannons it is found to be less corrosive on the bore of the gun than guncotton, which is a more highly nitrated form.

It is used in making smokeless powder. It is colloided with volatile solvents with or without addition of some non-volatile solvents in order to convert it to *smokeless powder*. In gelatinized form it is used as the basis of all propellants, particularly cannon powder or in mixture with *guncotton* in smaller caliber guns; or mixed with nitroglycerin in double-base powders.

Synonym: Pyro.

Also see: Cellulose Nitrate, and Nitro-cellulose.

Pyrogen – A rocket ignition system containing a solid propellant gram as its main ignition material.

Pyrophoric Article – An article that may contain both an explosive substance or component and a *pyrophoric substance.*

Pyrophoric Substance – The substance that is capable of spontaneous ignition when exposed to air.

Pyro Powder – Straight nitrocellulose powder; smokeless propelling charge consisting of a nitrocellulose that has a smaller nitrogen content than guncotton; single-base propellant.

Pyrotechnic – The term 'pyrotechnic' has been derived from the Greek words 'pyr' (fire) and 'techne'. According to some, 'pyrotechnic' means 'science of fire'. The term embraces all chemical devices and materials whose purpose is to burn and produce an effect by explosion, fire, light, heat, gas or smoke or a combination of these as a result of non-detonative self-sustaining exothermic chemical reactions.

Pyrotechnics are, in effect, any combustible or explosive compositions or manufactured articles. In other words, pyrotechnics is an explosive device that burns with colored flames.

Pyrotechnic substances are included in the definition of 'explosive' even when they do not produce gases. Pyrotechnics are low explosives because of their low rates of combustion.

'*Fireworks*' are pyrotechnic devices used for entertainment. However, pyrotechnics are commonly referred to as fireworks. According to some opinion, 'pyrotechnic' is the Greek word for 'fireworks'.

Pyrotechnics are used to illuminate areas of interest (such as targets for gunfire), to send signals by visual means (such as color), to simulate other weapons or activities. They can also be used as ignition elements for certain types of weapons.

Pyrotechnic devices include airbag inflators, delay compositions, flares, fireworks, fuel pellets for field stoves and other heating units, illuminating compositions, illuminating projectile or star shell used to illuminate targets for gunfire, military incendiaries, quick-release devices used for emergency exit systems in aircraft, smoke grenades, and solid rockets. Although encased in a projectile body of standard external shape and fired from a standard rifled gun, a star shell is actually a pyrotechnic device.

Pyrotechnic Composition – A mixture of chemical substances, which on burning and without explosion produces visible or brilliant displays or bright lights, or whistles, or motions.

Pyroxylin – *Nitrocellulose* containing 8-12% nitrogen.

Also see: *Nitrocellulose.*

Quantity-Distance Table – A table listing minimum recommended distances from explosive magazines of various weights to a specific location.

Quick Fuse, Super – Super Quick Fuse: A fuse that functions at the quickest possible time upon impact of the missile with the target.

Synonym: Instantaneous fuse.

Quick Match – *Black match* that is encased in a loose-fitting paper or plastic sheath to make it burn extremely rapidly is termed quick match. It consists of cotton wicking impregnated with gunpowder and covered with a loose paper piping.

Quick match is used for conveying fire to the combustible portion of pyrotechnical devices such as aerial shells. It is distinguished from a fuse by the fact that its effect is almost instantaneous, whereas a fuse burns at a comparatively slow and exact rate. However, quick match is also known as instantaneous fuse. It is used for simultaneous ignition of a number of pyrotechnic devices, such as lances, in a ground display piece.

As almost every piece of firework requires a match for lighting, and since lance-work and exhibition pieces are absolutely dependent on a good match for their successful operation, it is essential to make this very necessary article as nearly perfect as possible.

Quincke Tube – The term 'Quincke Tube' means filtering device for acoustic waves.

Radiological Dispersal Device – A radiological weapon that combines radioactive material with conventional explosives. It is a synonym of 'dirty bomb'. Although an RDD is designed to disperse radioactive material over a large area, the conventional explosive would likely have more immediate lethal effect than the radioactive material.

Acronym: RDD.

Railway Fog Signal – A warning device, designed to be placed on the rail, containing a composition that explodes with a loud report when the device is crushed.

Synonyms: Railway Fog Signal, Railway Torpedo, and Railway Track Explosive Signal.

Railway Torpedo – Synonym of *Railway Fog Signal.*

Railway Track Explosive Signal – Synonym of *Railway Fog Signal.*

Rapid Phase Transition – Rapid phase transition means the rapid change of state of a substance that may produce a blast wave and missiles.

Range, Explosive – Explosive Range: The difference between the lower and higher explosion or flammability limits, expressed in terms of percentage of vapor or gas in air by volume, is known as the "explosion range", also often referred to as the "flammability range." For example, the lower limit of flammability of gasoline at ordinary ambient temperatures is approximately 1.3 percent vapor in air by volume, while the upper limit of flammability is about 6. By difference, the explosive or flammability range of gasoline is therefore 4.7.

Reaction, Explosive – Explosive Reaction: In most explosives (but not all), the process by which large quantities of heat and gas are produced is oxidation, i.e. combustion in which the oxidizer is provided by the explosive, unlike normal combustion where the oxidizer is atmospheric oxygen. Once this oxidation process in the explosive has been initiated, it will proceed without any additional energy or material being required, until the explosive has been fully consumed. In some explosives, such as black powder, the fuels (carbon and sulphur) are present as separate ingredients mixed intimately with the oxidizer (potassium nitrate). In other explosives the fuel and oxidizer, are present within the same molecule, e.g. trinitrotoluene (TNT), where the fuels are carbon and hydrogen atoms, and the oxidizers are nitro (NO_2) groups.

Also see: *Initiation of Explosive Reaction.*

Rarefied Explosion – Rarefied explosions arise mainly from the ignition of admixtures of flammable gas/vapor, or extremely fine liquid or solid particles with air. Such an explosion is characterized by a much lower energy density than a *dense explosion*. The overpressures generated by rarefied explosions are thus much lower, being of the order of 10 bars. Such explosions lack brisance and do not give rise to craters. Rarefied explosion is a *non-seated explosion.*

Rarefied explosions may be subdivided into vapor explosions and particulate explosions. Rarefied explosions may be either confined or unconfined. The vast majority of rarefied explosions are confined.

Also see: *Family Tree of Explosions*; and *Aerosol Explosion*, *Dust Explosion*, *Particulate Explosion, Successive Explosion.*

RDD – Acronym for *Radiological Dispersal Device.*

RDF – Acronym for Reinforced Detonating Fuse. Most frequently applied to reinforced MDF.

RDX – An acronym for Research Department Explosive. The acronym is also translated as Royal Demolition Explosive. The secondary high explosive compound chemically named cyclotrimethylenetrinitramine, the empirical formula being $C_3H_6N_6O_6$.The structural formula is as below:

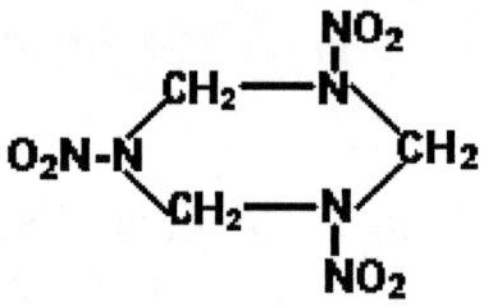

NO_2
CH_2—N
O_2N-N CH_2
CH_2—N
NO_2

RDX

It is also known as 'Cyclonite', 'Hexogen' and 'T4'. The U.S. military named this explosive 'RDX'; the name 'Cyclonite' was given to this explosive by Clarence J. Bain because of its cyclic structure and its cyclonic nature; the Germans call it Hexogen, and the Italians T4.

RDX is a white crystalline powder with molecular weight of 222.1, density 1.82 g/cm^3, oxygen balance -21.6%, nitrogen content 37.84%, detonation velocity 817 meters per (26,800 feet) per second.

RDX exhibits very high brisance (shattering power). It is a handy and versatile material for bringing about all kinds of destruction. RDX has a high degree of stability in storage and is considered the most powerful high explosive. Today, Cyclonite is probably the most important high brisance explosive. Its brisant power is high because of its high density and high detonation velocity. It is relatively insensitive in comparison to PETN that has similar strength. When combined with proper additives, such as plasticizer, it is relatively stable and insensitive to shock. RDX is very stable, insoluble in water, sparingly soluble in alcohol, ether and benzene, and soluble in acetone. RDX is the main ingredient in plastic explosives, and used extensively by the military.

Compositions in which RDX is melted with wax are called Composition A.

Compositions in which RDX is mixed with TNT are called Composition B.

Compositions in which RDX is blended with a non-explosive plasticizer are called Composition C.

An explosive substance used in the manufacture of Compositions B, C-3 and C-4. Composition B is useful as a cast primer.

RDX was not widely used until World War II. It is usually used in mixtures with other explosives, oils, or waxes. It is used in the manufacture of Compositions A, B, and C. It has found its use in the manufacture of boosters, detonating cord, detonators, and placed in a plastic binder. Pure RDX is used in press-loaded projectiles. Cast loading is accomplished by blending RDX with a relatively low melting point substance. RDX is also used as a base charge in detonators and in blasting caps.

It is commonly used as a booster in explosive trains or as a main bursting charge. It is stable in storage, and when combined with proper additives, may be cast or press loaded. It may be initiated by lead azide or mercury fulminate. It may be used as a booster or main bursting charge. When combined with proper additives, such as plasticizer, it is relatively stable and insensitive to shock.

Synonyms: 'Cyclonite', 'Hexogen' and 'T4'.

Reaction Time – The time between the ignition of the fuse head and the explosion of the detonator, in firing an electric detonator.

Realgar – Natural red *Arsenic disulphide*, As_2S_2.

Red H – Non-gelatinous permissible explosive, used in coalmines.

Redox Reaction – Acronym for oxidation-reduction reaction. It is a chemical reaction in which an oxidizing agent is reduced and a reducing agent is oxidized, thus involving the transfer of electrons from one atom, ion, or molecule to another. The 'redox potential' is the potential required in a cell to produce oxidation at the anode and reduction at the cathode. This potential is measured relative to a standard hydrogen electrode, which is taken as zero.

Reducing Agent – An element or compound that is capable of removing oxygen or adding hydrogen, or one that is capable of giving electrons to an atom or group of atoms.

Reinforced Ordinary Detonator – Synonym of *Plain Detonator.*

Reflected Overpressure – When a rigid object is perpendicular to the advance of the blast wave (facing), the object reflects and diffracts the wave. Due to this reflection, the object experiences an effective overpressure, which is called the reflected overpressure. The reflected overpressure is of at least twice the side-on overpressure.

Relative Bulk Strength (RBS) – See under *Explosive Energy.*

Relative Weight Strength (RWS) – See under *Explosive Energy.*

Relay, Explosive – Explosive Relay: An element of a fuse explosive train, which augments an outside and otherwise inadequate output of a prior explosive component so as to reliably initiate a succeeding train component. Relays, in general, contain a small single explosive charge such as lead azide and are not usually employed to initiate high explosive charges.

Relay, Detonating – Detonating Relay: It is a device that consists of metal capsules with a diameter almost identical to that of a detonator. It is open at both ends and contains a delay element, similar to that of an electric delay detonator, between two small charges of detonating explosive. They are so constructed as to allow easy connection to the detonating cord. In practice, the cord is cut and the ends are inserted into either end of the relay and crimped with a special crimping tool. When the detonation reaches the relay it is interrupted, the delay element operates introducing a delay, and reinitiates the cord at the other end. As and when a sequential firing is required, detonating relays may be inserted in the firing line.

Release Device, Explosive – Explosive Release Device: A device containing a small electrically initiated explosive charge, and which is comprised of rods or links fitted with means for mechanical attachment to other equipment or apparatus to be released or severed.

Remote Operation – Remote operation means an operation performed in a manner that will protect personnel in the event of an accidental explosion. The object of a remote operation is to maintain distance, shielding, barricades, or a combination thereof.

Remover – An explosive telescoping device designed to remove a canopy from an aircraft.

Report – A pyrotechnical term that denotes the loud boom that is created when a salute ignites.

Research Department Explosive – See *RDX.*

Rest House – Synonym of *Service Magazine.*

Revolver – A type of handgun having multiple chambers, each of which revolves into position to fire a cartridge. Generally the number of chambers in a revolver is six. However, revolvers with five, seven, and nine chambers are not unusual.

Rifle – A firearm that has rifling (spiral grooves) in the bore in order to give a spin to the projectile so that it will have a greater accuracy of fire and longer range. A rifle is designed to fire from the shoulder and fire only a single projectile at a time, as opposed to a shotgun, which can throw many small projectiles (shot) at the same time.

Rifle Grenade – Grenades that are projected from a launching device attached to the muzzle of a rifle or carbine, and requiring a special cartridge.

Rim-fire Cartridge – A cartridge in which the primer (fulminate) is located in a rim surrounding its base. In this type of cartridge, primer (fulminate) is integral to the shell casing. When the firing pin strikes, the pin pinches the rim against the chamber and causes the rim to explode and ignite the powder.

Ripping – In coal mining, the removal of stone after recovery of coal to draw up a road of normal size.

Risk – A measure of the combination of the probability and consequences of the hazards of an operation, expressed in qualitative or quantitative terms.

Rivet, Explosive – Explosive Rivet: A metallic rivet containing an explosive composition. A rivet containing an explosive charge that is exploded either by touching head with a heated iron or placing it in a high-frequency electromagnetic field.

Roburite – Name of a smokeless and flameless mining explosive, in the compounding of which chlorodinitrobenzene is employed, in admixture with *ammonium nitrate*.

Rocket – 1. A rocket is any self-propelled object designed to travel above the surface of the earth. The term 'rocket' includes any rocket or missile, military or civil, with or without guidance. In other words, a rocket may be defined as a missile containing combustibles, independent of atmospheric oxygen, which on being ignited, liberate gases producing thrust. 2. A pyrotechnical term which means an artificial firework consisting of a cylindrical case of paper or metal filled with a composition of combustible ingredients, as niter, charcoal, and sulphur, and fastened to a guiding stick. The rocket is projected through the air by the force arising from the expansion of the gases liberated by combustion of the composition. Rockets are used as projectiles for various purposes, for signals, and also for pyrotechnic display.

Rocket, Congreve – Congreve Rocket: A powerful form of rocket for use in war. It was invented by Sir William Congreve and named after him. The rocket is designed for use either in the field or for bombardment. In the case of field use, it is armed with shells or case shot. In the latter case, it is armed with a combustible material enclosed in a metallic case. It is inextinguishable when kindled, and scatters its fire on every side.

Rocket Motor – The contrivance usually contains a charge of solid propellant placed in a metal cylinder fitted with one or more nozzles, which is designed for propelling a rocket, missile, etc.

Rocket Motor Igniter – A rocket motor igniter is an explosive device designed and used to ignite the propelling charge in a rocket motor. It consists of an electric igniter and a fast burning composition assembled as a unit.

Rocket Propellant – 'Rocket Propellant', often shortened to 'propellant' is any agent used for consumption or combustion in a rocket and from which the rocket derives its thrust, such as a fuel oxidizer, additive, catalyst, or any compound or mixture of these.

Synonym: *Propellant.*

Rocket with Bursting Charge – A rocket that consists of a rocket motor and an explosive warhead.

Rocket with Inert Head – Rockets fitted with motors that are normally ignited by electric primers or electric squibs.

ROD – Acronym for *Reinforced Ordinary Detonator.*

Roman Candle – A paper tube firework from which a series of colored balls of fire are projected.

Rotational Firing – Rotational firing means the delay blasting system used so that the detonating explosives will successively displace the burden into the void created by previously detonated explosives in holes, which were fired at an earlier delay period.

Round – One shot fired by a weapon. The term is also used to mean all the parts that make up the ammunition necessary in firing one shot.

Round, Complete – Complete Round: Synonym of *Round.*

Round, Fixed – Fixed Round means round of fixed ammunition.

Royal Demolition Explosive – See *RDX.*

Rule, Kistiakowsky-Wilson – See *Kistiakowsky-Wilson Rule.*

Russian A 91 – *RDX* with a small amount of paraffin stabilizer added.

Rye Meal – A vegetable meal used in the manufacture of *dynamites* as an absorbent for nitroglycerine.

Safe And Arm – The term 'safe and arm' as used in explosive trains means a device for interrupting (safing) and aligning (arming) an explosive train.

Safe, Intrinsically – See *Intrinsically Safe.*

Safe Load – Safe load means the stress to which a given substance may be subjected in common working practice.

Synonym: *Safe Stress.*

Safe Stress – Synonym of *Safe Load.*

Safety Cartridge – A cartridge for small arms having a diameter not exceeding 2.5 centimeters (1 inch), the case of which can be extracted from small arms after firing and which is closed so as to prevent any explosion in one cartridge being communicated to other cartridges.

Safety Celluloid – Safety celluloid is one that is not highly inflammable, is not easily ignited, and does not burn rapidly and fiercely unlike normal celluloid made from *cellulose nitrate.* In the case of safety celluloid, stringent precautions for storage and handling are not required, as opposed to normal celluloid. Safety celluloid is usually made from cellulose acetate; that means the cellulose is treated with acetic acid instead of nitric acid. It burns in the manner of paper rather than the fierce and rapid way of celluloid. It is used for making cinematographic film, photographic film, transparent wrapping papers, etc.

Safety Circuit for an Electric Detonator Element – A device to prevent improper initiation of the electrical detonator element. The safety circuit for an electric detonator element is connected to two electrical leads, and has a subsequently arranged detonable charge. The safety circuit includes an electrical component connected in parallel to the detonator element between the two leads, the electrical component having an initial condition wherein the electrical component is at least substantially electrically non-conductive. The electrical component is responsive to a signal having a voltage value above a maximum ignition voltage value necessary to enable triggering of a detonator element. The electrical component provides a short circuit path in parallel to the detonator element so as to prevent improper initiation of the electrical detonator element.

Safety Distance – Synonym of *Safety Zone.*

Safety Explosive – 1. An explosive that requires a powerful initial impulse and therefore may be handled safely under ordinary conditions is often referred to as safety explosive.

2. The term Safety Explosive is now obsolete. The explosive in which the explosion temperature is relatively low and enduring only a short time; the flame of explosion is relatively short and of short duration.

The use of ordinary high explosives in coal mining is hazardous because of the danger of igniting gases or suspended coal dust that may be present underground. Special types of explosives have been developed to carry out blasting under such conditions, in order to minimize the danger of explosions or fires by producing flames that last for a very short time and are relatively cool. The special types of explosives have been termed safely explosives.

This type of explosive originally contained ammonium nitrate and charcoal. Numerous safety explosives of various combinations have been on the market, chiefly mixtures of ammonium nitrate with other ingredients such as sodium nitrate, nitroglycerin, nitrocellulose, nitro starch, carbonaceous material, sodium chloride, and calcium carbonate.

Another type of blasting charge, a liquid carbon dioxide blasting charge, which may be obsolete and was once used in mining, produces no flame whatsoever. This blasting charge is a cylinder of liquid carbon dioxide that can be converted into gas almost instantaneously by an internal chemical heating element. One end of the cylinder contains a breakable seal through which the gas can expand. The carbon dioxide charge is not a true explosive and instead of creating heat, it absorbs heat. Liquid carbon dioxide blasting charge has an additional advantage that the force of the explosion can be directed at the base of the borehole in which the charge is placed, thus lessening the shattering of the coal.

Synonyms: Permissible Explosive (USA), and Permitted Explosive (UK).

Also see *Liquid Carbon Dioxide Explosive*.

Safety Factor – It is the ultimate strength of a substance divided by actual unit stress on a sectional area.

Synonym: Factor of safety.

Safety Film – Photographic film made of cellulose acetate.

Also see: *Safety Sampson Film.*

Safety Fuse – It is a fuse for igniting charges of other explosives safely. Initiation of an explosive charge has to be done remotely, and one method of achieving this is the use of a safety fuse. It provides a dependable means of conveying flame to the charge.

The invention of the safety fuse is the first important development in the history of explosives, followed by the invention of blasting cap. William Bickford, a British leather merchant devised the safety fuse in 1831, which for the first time enabled safe, accurately timed detonations. Before the invention of safety fuse, accidents in the mines were frequent, owing to the deficiencies inherent in the quill fuse.

Safety fuse consists of a core of fine-grained black powder surrounded by a flexible woven fabric with one or more protective outer coverings. Fire or flame is conveyed by it at a continuous and uniform rate from the point of ignition to the point of use. The minimum length of safety fuse to be used in blasting is not less than 75 centimeters (30 inches).

It does not contain its own means of ignition. It is of such strength and construction, and contains the explosive (black powder) in such quantity, that the burning of such fuse would not communicate laterally with other like fuses.

Safety fuse has the ability to ignite black powder in a blast hole without the need for a detonator. However, to initiate high explosives, a safety fuse must be used in conjunction with a fuse detonator (plain detonator). Thus safety fuse may be defined as a device by which a flame is communicated to the detonator or charge either after an interval of time, or by an operation conducted at a distance. In either case the shot firer is not exposed to the effects of the explosion.

When ignited, safety fuse merely burns slowly and does not explode. The burning rate of safety fuse is nominally 100 seconds per meter (30 seconds per foot). This may vary between 90 and 110 seconds per meter (27 and 33 seconds per foot). Because of this characteristic of unreliable timing, the use to safety fuse has become limited in commercial blasting because an initiation system is expected to transfer the detonation signal from hole to hole at a precise time. Safety fuse was usually used where sources of extraneous electricity made the use of electric detonators dangerous.

Safety Gasoline – Hydrogenated gasoline with higher flash point than ordinary gasoline.

Safety Head – The term 'safety head' relates to pressure vessels and means a disk, usually of metal, which will burst at a certain pressure, relieving a system of excess pressure.

Safety Lamp – An oil-lamp that will not ignite inflammable gases, such as methane (*fire-damp*). It is comprised of a cylinder of wire gauze acting as a chimney; the heat of the flame is conducted away by the gauze, and while the fire-damp will burn inside the gauze, the temperature of the gauze does not rise sufficiently high to ignite the gas outside.

Synonym: *Davy Lamp.*

Safety-lamp Mine – Safety lamp mine means a gassy mine, implying possible presence of inflammable gases, such as methane (*fire- damp*).

Safety-lamp Oil – Safety-lamp oil is a mixture of 60% sperm oil and 40% mineral seal oil.

Safety Match – A match designed to ignite only when rubbed on a specially prepared surface. A safety match is ignited by the generation of heat on the striking surface of the box, the coating of which consists mainly of red phosphorus, ground glass, and glue. The match head contains potassium chlorate, mixed with a number of other ingredients including sulphur, antimony sulfide and glue. Antimony sulfide is used in the heads as a flame-producing agent. When the match is rubbed on the striker, the glue coatings on the two reactants are broken and the reactants come into intimate contact and immediately react.

Compare: *SAW Match*, *Strike Anywhere Match.*

Safety Mechanism, Igniter – See *Igniter Safety Mechanism.*

Safety Plug – Safety plug means the fusible metal plug set in the shell of a boiler in order to release steam from the boiler when the plug reaches a predetermined temperature due to excessive pressure in the boiler.

Safety Shoe – A shoe with a sole of material incapable of sparking; used for work near explosives or flammables.

Safety Squib – The term 'safety squib' means a small paper tube containing a small quantity of black powder.

Safety Standard – A set of prescribed precautions relating to the practices to be observed to ensure safety in the manufacture, transportation, storage, handling and use of explosives.

A common sense approach in the handling of explosives in any form is a must. Safety in the manufacture, transportation, storage, handling and use of explosives based on good management rather than good luck.

Safety Tube – See *Tube, Safety.*

Safety Valve – See *Valve, Safety.*

Safety Wire – See *Wire, Safety.*

Safety Zone – See *Zone, Safety.*

SAFEX International – Abbreviation for Safety in Explosives. Cooperation between explosives manufacturing companies and organizations began in 1954, and continues to date in the name of Safety in Explosives (SAFEX International), a worldwide safety organization created to pool safety and accident information worldwide.

Safe Zone – Synonym of *Cold Zone.*

Salt-Bath Explosion – See *Explosion, Salt-Bath.*

Saltpeter – The chemical compound potassium nitrate, KNO_3 is known as saltpeter. It occurs as colorless prismatic crystals or as a white powder. When heated, it decomposes to release oxygen. Saltpeter has been used in *gunpowder* manufacture since about the 12th century. It is also used in explosives, fireworks, matches, and fertilizers, and as a food preservative.

Synonyms: Niter, and Potassium Nitrate.

Saltpeter, Chile – See *Chile Saltpeter.*

Salutes – Devices that contain powdered aluminum or an aluminum/magnesium alloy which, when ignited, can result in a violent explosion and flash.

SAW Match – See *Match, SAW.*

SCID – Acronym for *Small Column Insulated Delay.*

Screening Smoke – The term 'screening smoke' denotes chemical agents which when burned, hydrolyzed or atomized produces an obscuring smoke. The objective of its use is to deny observation and reduce effectiveness of aimed fire.

Seated Explosion – An explosion in which the point of origin of the explosion, or the seat of the explosion can be ascertained is known as a seated explosion. The crater or the area of greatest damage that is caused by explosion (also called epicenter) is generally one of the characteristics of dense explosions. Seated explosions are generally the result of the initiation of solid or liquid chemical explosives. However, the area where a physical explosion occurs will leave a visible area to indicate the location of the epicenter.

Compare: Non-seated Explosion.

Secondary Blasting – A process of breaking, by the use of explosives, boulders resulting from a primary blast that are too large for immediate handling. The oversize material resulting from an initial blast is reduced to the dimension required for handling by the process of secondary blasting, which includes *mud-capping* and *block holing.*

Secondary Device – A device that is placed in order to maim or destroy first-response personnel. It detonates after the main charge has exploded and first-response personnel have gathered on the scene.

Secondary Explosive – 'Secondary high explosive' is usually shortened to 'secondary explosive'. As the term 'secondary' implies, it is a high explosive so insensitive to heat or shock that it does not detonate (unlike a primary explosive) but is detonated by suitable devices.

Since a secondary explosive is relatively insensitive, a great amount of energy is needed to initiate decomposition. Some secondary explosives are insensitive enough that they can be lit with a match and they simply burn like wood. The initiation of secondary explosives generally requires the shock wave energy from a primary explosive. A high explosive is usually initiated by a primary explosive or by an exploding bridge wire. Hence it requires a detonator to explode.

Examples of secondary explosives are: Nitroglycerin, Dynamites (and other nitro-glycerin based explosives), TNT, PETN, RDX, HNS, HMX, PBX, Amatol, Tetryl, Picric acid, Gelignite, Nitro starch, Cellulose nitrate (guncotton), Urea nitrate.

Secondary explosives have the highest energy outputs of any explosive and are used for blasting, demolition, or fragmentation type applications where a high-energy shock wave is desirable. Secondary explosives are used as a main charge and also as a primer, or booster, to initiate tertiary explosives.

Also see: *Detonating explosive, Family Tree of Explosives, High explosive,* and *Primary explosive.*

Synonym: *Secondary High Explosive.*

Secondary Fragmentation – Fragments produced by an explosive device that is made up of the target materials or other materials other than those directly resulting from the device itself.

Secondary High Explosive – Synonym of *Secondary Explosive.*

Secondary Wave – A wave created when the air around the blast rushes in to fill the vacuum created during the original explosion.

Seismic Detonator – Detonators made for use in geophysical surveying applications. These are a specific type of instantaneous electric detonator designed to function in less than one millisecond following application of the recommended level of current and are designed for rugged field usage. Seismic detonators typically incorporate a time break function in which the circuit is interrupted only at the moment of detonation, providing a highly accurate zero time on the seismic recording trace.

Seismic Explosive – Explosives made for use in seismic prospecting. The seismic explosives used for geophysical surveying applications for determining the depth of geological formations, and for locating minerals, ores and oil- and water-bearing strata are necessarily especially compounded to withstand the severe conditions of seismic prospecting.

Reliable and complete propagation under high hydrostatic pressures, full velocity of detonation under these conditions, very good water resistance to be capable of remaining in water even for weeks, rigid packaging, and long shelf life are the especial characteristics of seismic explosives.

Self-contained Explosive – The compound that contains the oxygen and the oxidizable elements carbon and hydrogen within the same molecule is a self-contained explosive. The organic nitrates (nitrocellulose, nitroglycerine, nitrostarch) contain the oxygen and the oxidizable elements carbon and hydrogen within the same molecule; they are self-contained explosives. The same is true for the aromatic nitro explosives (trinitrotoluene, trinitrophenol).

Also see: *Explosophore Group.*

Self-ignition Point – Synonym of *Auto-ignition Point.*

Semi-conductive Bridge Detonator – In this technology, the bridge wire is replaced by a microchip containing a semi-conductive bridge, which flashes upon the application of a low current. This design, when combined with microelectronic circuits, can be used to produce electronic delay detonators.

Semi-conductive Hose – A hose used for pneumatic conveying of explosive materials having an electrical resistance high enough to limit flow of stray electric currents to safe level yet not so high as to prevent drainage of static electric charges to ground. Hose of not more than 2 mega ohms resistance over its entire length and of not less than 5,000 ohms per 30 centimeters meets the requirement.

Semi-Gelatin – Explosive containing nitroglycerine gelled with nitrocellulose, but in quantity insufficient to fill the voids between the salt and combustible particles and thereby produce a gelatin.

Semi-Gelatin Dynamite – See *Dynamite, Semi-Gelatin.*

Semtec – Synonym of *Semtex.*

Semtex – Trade name of a *plastic explosive* (*plastic bonded explosive-PBX)* that is a varying blend of *PETN* and *RDX* with styrene-butadiene copolymer as a plasticizer. Manufactured in Czechoslovakia while under communist rule, much of this product was circulated and has been utilized by various terrorist organizations. Usually dyed orange.

Synonym: Semtec.

Sensitiveness – Sensitiveness is a measure of an explosive's ability to propagate a detonation. It is in effect cartridge-to-cartridge propagating ability under certain test conditions. It is expressed as the distance through air at which a primed half-cartridge (donor) will detonate an unprimed half-cartridge (receptor).

Sensitivity – Sensitivity is a physical characteristic of an explosive classifying its ability to be initiated upon receiving an external impulse such as impact, shock, flame, friction, or other influences that can cause explosive decomposition. In short, the ease with which the chemical reaction can be initiated is known as sensitivity.

Sensitivity, Maximum Gap – See *Maximum Gap Sensitivity.*

Sensitizer – An ingredient used in explosive compounds to promote greater ease in initiation or propagation of the detonation reaction.

Separated Ammunition – Separated ammunition is characterized by the arrangement of the propelling charge and the projectile for loading into the gun. The propelling charge, contained in a primed cartridge case that is sealed with a closing plug, and the projectile are loaded into the gun in one operation.

Separated ammunition is usually used where the fixed ammunition is too large to handle.

Separate-Loading Ammunition – Ammunition in which the propelling charge, primer and projectile are not held together in a shell case, as in fixed ammunition. These are loaded into the gun separately. No cartridge case is utilized in this type of ammunition.

Separating Burst – Method of ejecting the contents of a projectile by means of a charge of propellant that breaks the projectile into two approximately equal parts, along a specially designed circumferential shear joint.

Sequential Blasting Machine – A series of condenser discharge blasting machines in a single unit, which can be activated at various accurate time intervals following the application of electrical current.

Series Blasting Circuit – An electric blasting circuit that provides one continuous path for the current through all detonators in the circuit.

Series in Parallel Blasting Circuit – An electric blasting circuit in which the ends of two or more series of electric detonators are connected across the firing line directly or through bus wire.

Serpent – A firework that moves in serpentine manner when ignited.

Service Magazine – An auxiliary building or suitable designated room, such as a vault, used for the intermediate storage of explosives not exceeding the minimum amount necessary for safe and efficient operation.

Synonym: *Rest House.*

Shaped Charge – Normally explosive force is released at 90-degree angles from the surface of an explosive. If the surface is cut or shaped the explosive forces can be focused directionally, and will produce a greater effect. This is known as a shaped charge. In other words, shaped charge may be defined as an explosive charge designed to produce specific effects by the inclusion of a re-entrant conical or V-shape, usually lined with metal.

It is in effect an explosive with a shaped cavity, specifically designed to produce a high velocity cutting or piercing jet of product reaction; usually lined with metal to create a jet of molten liner material.

Synonym: Cavity Charge.

Shaped Charge, Flexible Linear – Flexible Linear Shaped Charge: A charge with a liner, specially shaped to produce a cutting jet. It is a linear version of shaped charge.

Shaped Charge, Linear – See *Linear Shaped Charge*.

Sheathed Explosive – The permissible/permitted explosive whose safety factor is further increased by sheathing it with the cooling agent, powdered sodium bicarbonate, which is used externally with little effect on the power or sensitiveness of the charge, is known as a 'sheathed explosive'.

The cooling agent sodium bicarbonate absorbs heat from the gases produced by the reaction, reducing their temperature. Moreover, sodium bicarbonate is decomposed by the heat and carbon dioxide - a fire extinguishing gas is evolved. A blanket of carbon dioxide is thus formed instantly between the hot gases from the explosion and any flammable gases that may be present.

The weight of a sheathed cartridge, as shown on the outside wrapper, is exclusive of sheathing. The ends of an explosive cartridge are not sheathed, because an inert layer between adjacent cartridges reduces sensitiveness to propagation without any compensating increase in safety.

Sheet Explosive – Sheet explosive is a flexible high explosive. It is practically a *plastic bonded explosive.* It is composed of an integral mixture of secondary high explosive, such as PETN or RDX, and a binder. This explosive remains flexible over a wide range of temperatures. It is waterproof and available in a variety of extruded shapes and in sheets and cords. These plastic bonded explosives have a very high brisance and detonating velocity.

Sheet explosives are designed for use as a cutting, breaching, or cratering charge, and especially for use against steel targets. The sheets of explosives may be quickly applied to irregular and curved surfaces, and are easily cut to any desired dimension.

Sheet explosives are known by many trade names, such as Deta sheet, Metabel, Series 1000 - PETN sheet explosive, and Series 2000 - RDX sheet explosive.

Shell – 1. Shell is a hollow projectile filled with explosive, or chemical or other material. Shell is contrast to shot, which is a solid projectile. 2. Ammunition consisting of a cylindrical metal casing containing an explosive charge and a projectile.

Shell, Aerial – See *Aerial Shell.*

Shell, High Explosive – See *High Explosive Shell.*

Shell, Illuminating – See *Illuminating Shell.*

Shelf life – The length of time of storage during which an explosive material retains adequate performance characteristics.

Shell, Smoke – See *Smoke Shell.*

Shell, Squashed-Head – See *Squashed-Head Shell.*

Shield, Blast – See *Blast Shield.*

Shielded Mild Detonating Cord – See *SMDC.*

Ship Distress Signal – See *Signal, Ship Distress.*

Also see: *Illuminating Ammunition.*

Shock Energy – The shattering force of an explosive caused by the detonation wave.

Shock Tube – See Tube, Shock.

Shock Tube Detonator – The term 'shock tube detonator' refers to a device that includes a length of shock tube.

Shock Tube System – A system for initiating caps in which the energy is transmitted to the cap by means of a shock wave inside a hollow plastic tube.

Shock Wave – See *Wave, Shock.*

Shooting, Coyote – Coyote Shooting: Synonym of *Coyote Blasting.*

Short Delay Blasting – The term is used to mean the practice of initiating individual explosive charges to detonate in successive intervals, in which the time difference between any two successive detonations is measured in milliseconds.

Short Delay Detonator – Short delay detonator is a delay detonator with time interval between individuals of the series of 25 or 50 milliseconds.

Short Delay Fuse – One that will burst a projectile on ricochet, preferably about 2 to 3 meters (6 to 10 feet) above ground. Some crater effect will be obtained on hard ground.

Shot – 1. A solid projectile for a cannon, without a bursting charge. 2. Pellets, small balls, or slugs in shotgun shells, and some other types of ammunition.

Shot Anchor – A device that anchors explosive charges in the borehole so that the charges will not be blown out by the detonation of other charges.

Shot Firer – That qualified person in charge of and responsible for the loading and firing of a blast. Shot firer is the person who actually fires a blast.

Synonyms: *Blaster, and Shooter.*

Compare: *Powder man.*

Shot Firing Operation – See *Operation, Shot Firing.*

Shot Hole – Shot Hole means a bored hole that is to be charged with explosive charges for blasting purposes.

Shrapnel – Artillery projectile that contains small lead balls that are propelled by a powder charge in the base, set off by a time fuse. Shrapnel has been replaced almost entirely by high-explosive shells. Wounds called shrapnel wounds usually are due to shell fragments rather than to shrapnel.

Shunt – The shorting together of the free ends of: (1) electric detonator leg wires, or (2) the wire ends of an electrical shorting device applied to the free ends of electric detonators by the manufacturer.

Signal – A pyrotechnic item designed to produce a sign to provide identification, location, warning, etc. Such a sign may be an illumination, smoke or sound.

Signal Cartridge – Cartridges designed to fir colored flares from "Very" pistols, etc.

Signal Device, Hand – See *Hand Signal Device.*

Signaling Smoke – Any type of smoke, but usually colored smoke from a hand or rifle grenade.

Signal, Railway Fog – *Railway Fog Signal.*

Signal, Railway Track Explosive – Railway Track Explosive Signals: A warning device, designed to be placed on the rail, containing a composition that explodes with a loud report when the device is crushed.

Synonyms: *Railway Fog Signal*, and *Railway Torpedo.*

Signal, Ship Distress – Ship Distress Signal: It contains pyrotechnic substances and is designed to produce signals by means of sound, flame or smoke or any combination thereof.

Also see: *Illuminating Ammunition.*

Signal, Smoke, with or without explosive sound unit – Smoke signal, with or without explosive sound unit. It contains pyrotechnic substances that produce colored smoke signals, and in addition may or may not produce audible signals.

Signal, Warning – See *Warning Signal.*

Sign, Explosive – See *Explosive Sign.*

Single-Base Propellant – A propellant whose principal active ingredient is a simple form of nitrocellulose powder; the single-base propellant does not contain nitroglycerine.

Single-Section Charge – Single-section charge means a propelling charge in separate-loading ammunition that is loaded into a single bag. Unlike a multi-section charge, it cannot be increased or reduced for changes of range.

Silver Acetylide – Silver acetylide, (Ag_2C_2) is a primary high explosive.

Also see: *Acetylide.*

Site, Blast – See *Blast Site.*

Skyrocket – 1.Skyrocket is a species of fireworks that sends a firework display high into the sky. It burns as it flies.

2. A device used to propel a lifesaving line or harpoon.

Slapper Detonator – Slapper detonator is an *EED* initiated by a rapid discharge of a high current through a metal foil. The expansion of the metal vapor causes a plastic or metal covering to be propelled across an air gap and detonate a high-density explosive pellet.

Slow-match – See *-Match, Slow.*

Slurry Explosive – Slurry is a type of explosives used in commercial blasting. A category of blasting agent. It is made by sensitizing thickened aqueous slurry of oxidizing salts. It consists of ammonium nitrate to which sodium nitrate is added, suspended in water thickened with natural gum so as to have a viscous consistency, usually sensitized with TNT; pentolite, smokeless powders, or aluminum may be added. Sometimes it is cross-linked to provide a gelatinous consistency; hence it may be called water gel. Such compositions are fluid when hot and can readily fill polythene bags or similar containers. On cooling, they thicken to slurry, which has good water resistance. It may be available as a packaged or bulk blasting agent.

SMDC – An acronym for Shielded Mild Detonating Cord: MDF contained in a small (180 inches diameter) steel tubing. Sometimes referred to as hardline CDF.

Small Arm – A gun, pistol, revolver or rifle having a caliber of up to about 25 millimeters (1 inch). Small arms are primarily designed to be held by the hand or to the shoulder and fired by one person.

Also see: *Musket*, *Pistol*, and *Rifle*.

Small Arm Ammunition – 'Small arm ammunition' means any cartridge or shell designed for use in a pistol, revolver, rifle, or shotgun held by the hand or to the shoulder, and includes ball, shot, or blank cartridges or shells, or cartridges for propellant-actuated power devices.

Small Arm Ammunition Primer – Small percussion-sensitive explosive charges encased in a cap or capsule and used to ignite propellant powder.

Small Arm Cartridge – Cartridges designed to be fired in arms, including machine guns, of caliber not larger than 19.1 mm. Except in the case of a blank cartridge, it consists of a cartridge case fitted with a primer, containing a propellant powder charge together with a projectile, which may be solid, tracer, lachrymatory or incendiary. It may be arranged in boxes, mounted on belts or placed in clips.

Small Arm Nitro-Compound – 'Small-arm nitro-compound' means nitro-compound adapted and intended exclusively for use in cartridges for small arms.

Small Arm Primer – The term 'small arms primer' means primers used for small arms ammunition. It is synonym of '*percussion cap'.*

Synonym: Percussion Cap.

Small Column Insulated Delay – See *Delay, Small Column Insulated.*

Smoke – Smoke is the air-borne suspension of solid particles from the products of deflagration or detonation. It is stated that dust particles that float in the air cannot be much larger than 0.01 millimeters in diameter. From a scientific point of view, smoke is defined as systems in which the solid particles exhibit Brownian movement when dispersed in a gas.

Smoke Bomb – Smoke Bomb is generally a pyrotechnic device that is used usually by terrorists in order to conceal the getaway route, or cause a diversion, or simply provide cover. Military smoke bombs exist, which employ powdered white phosphorus or titanium compounds.

Smoke, Colored – Colored Smoke: An aerosol of special dyestuffs dispersed by pyrotechnical reaction by explosion. Used for signaling and spotting.

Smokeless Gunpowder – See *Smokeless Powder.*

Smokeless Powder – Smokeless powder is a nitrocellulose-based powder whose color is light brown to black. Paul Vieille developed it in 1884. Nitrocellulose in its natural fibrous form has too high a rate of combustion to be suitable for a propellant. In order to manufacture nitrocellulose powder, the nitrocellulose in its natural fibrous form is dissolved in a mixture of alcohol and ether to form a gelatinous mass. The gelatinous mass is then rolled into sheets, cut into small sized flakes and dried. The finished product has the required propelling power. Thus the smokeless power is obtained.

Smokeless powder is not entirely smokeless nor is it a powder.

Smokeless powder is a low explosive that burns rapidly with little or no smoke. It is very sensitive to friction, heat and shock.

Smokeless powder has now universally replaced black powder as a propellant. It is used in single-base powders and double-base powders. It is used as a propellant in cannon, mortar rockets, propellant-actuated power devices, small arms ammunition, etc. It is used in improved military rifle powder, sporting powder and ball powder.

Synonym: *Smokeless Propellant.*

Also see: *Cellulose Nitrate*, and *Nitrocellulose.*

Smokeless Propellant – Synonym of *Smokeless Powder.*

Smoke Projectile – A projectile containing a smoke-producing chemical agent that is released on impact or burst.

Synonym: *Smoke Shell.*

Smoke, Screening – See *Screening Smoke.*

Smoke Shell – Any projectile containing a smoke-producing chemical agent that is released on impact or burst. The smoke may be white or colored.

Synonym: *Smoke Projectile.*

Smoke Signal with or without Explosive Sound Unit – See *Signal, smoke, with or without explosive sound unit.*

Smoke, Signaling – See *Signaling Smoke.*

Snapper – See F*irecracker.*

Soda Powder – *Black powder* made with sodium nitrate instead of potassium nitrate is sometimes called soda powder. It was available by 1857. Soda powder was made in several varieties; for example, one named Mammoth Powder that was designed for large artillery. Its use as a propellant was limited because both more modern artillery and guncotton became available at about the same time. As it turned out, soda powder was mostly used for blasting powder.

Also see: *Black Powder.*

Sodatol Booster – *Pentolite boosters* or *Composition B boosters* that contain amounts of sodium nitrate are called sodatol boosters.

Sodium Azide – Sodium azide, NaN_3, is used as a propellant in car *air bags.*

Also see: *Azides.*

Sodium Hypophosphite – Sodium hypophosphite may detonate if heated. It has no known explosive uses at the present time. It must be kept cool, and stored in a cool, ventilated place, away from acute fire hazards. It may be disposed of by dissolving it in water.

Sodium Nitrate – Sodium nitrate, $NaNO_3$ is an oxidizer used in dynamites and sometimes in blasting agents.

Sofar Bomb – A sound-producing bomb designed to detonate at a given depth under water.

Solid Coal – Implies coal that is being worked without the provision of a free face by undercutting or similar means.

Solid Propellant – Specifically, a rocket propellant in solid form, usually containing both fuel and oxidizer, combined or mixed, and formed into a monolithic (not powdered or granulated) grain.

Sonic Boom – A sonic boom is a *shock wave* produced by an object moving through the air at supersonic speed, i.e. faster than the speed of sound. An object, such as an airplane, moving through the air generates sound. When the speed of the object exceeds the speed of sound, the object forces the sound ahead of itself faster than the speed at which the sound would ordinarily travel. The piled-up sound takes the form of a violent shock wave propagating behind the object.

Sounding Devices, Explosive – Explosive Sounding Devices: Devices containing an explosive charge and a means of initiation. The devices function when they hit the seabed after having dropped.

Spalling – In explosive technology, spalling implies the breaking off of a scab of material from a free face as a result of the reflection of shock waves.

Spark Arrestor – Spark arrestor means any device, assembly or method of a mechanical, centrifugal, cooling or other type and size designed and used for the retrenching or quenching of sparks in exhaust pipes from internal combustion engines.

Sparkler – A firework that burns slowly and throws out a shower of sparks.

Special Gelatin – See *Gelatin, Special.*

Special Gelatin Dynamite – See *Ammonia Gelatin Dynamite*, the Synonym of Special Gelatin Dynamite.

Spin – In fuse manufacture, to wind on a spiral of textile yarns.

Splinter Bomb – The Infamous German SD-2 Splitterbombe, used during World War II. It is almost identical to the U.S.-made M-83 anti-personnel munition, commonly known as *Butterfly Bomb*, used during the 1950s and after.

Also see: *Cluster Bomb.*

Splitter Cord – See *Ignitacord.*

Spontaneous Combustion – Spontaneous combustion occurs as a result of the heat generated by the reacting substances themselves. It occurs when certain vegetable oils or paints oxidize under conditions that permit the heat of oxidization to be retained and to build up to the point where something is set on fire. The usual case is an oil-soaked rag.

Apart from this, the absorption of oxygen by certain animal oils and fats, and the 'absorption' of oxygen by certain substances, such as pyrites present in coal, may lead to an increase in temperature so as to bring forth spontaneous combustion. When hot glycerin is poured onto crystals of potassium permanganate, the substances soon begin to fume and burst into flame.

Synonym: *Spontaneous Ignition.*

Spontaneous Explosion – Potassium chlorate causes spontaneous explosion. This is due to the fact that its acid component, chloric acid, is an unstable acid and is easily decomposed. Consequently, a slight rise in temperature is sometimes sufficient to bring about an explosion. The tendency of potassium chlorate to explode is very strong in the presence of sulfur, sulfides and sulfates which sometimes release minute quantities of sulfuric acid. For this reason, compositions containing both chlorates and sulfates are strictly avoided. Other chlorates also act in a similar way.

Spontaneous Ignition – Synonym of *Spontaneous Combustion.*

Sporting Cartridge – See *Sporting Cartridge.*

Sporting Powder – As used in sporting ammunition, sporting powder consists of nitroglycerine and/or nitrates in combination with the nitrocellulose base. It is similar to that of double-base powders. It is more readily ignitable and has a higher rate of burning than straight nitrocellulose powder.

Spray – Fragments of a bursting shell. The nose, side and base sprays are the fragments thrown forward, sideways and rearward respectively.

Sprengel Explosive – Sprengel explosive is a mixture of nitrobenzene and red fuming nitric acid, generally in proportion 28:72. This type of explosive is very difficult and awkward to handle, so its use has been discontinued. It is a powerful and inexpensive explosive, and would have many uses. The components of this explosive have to be mixed in glass shortly before the explosive is used. This requires preparation and equipment not always available at the site of the explosion.

Springing – Springing is the creation of a pocket in the bottom of a drill hole by use of a moderate quantity of explosives in order that larger quantities of explosives may be inserted.

Squib – The word 'squib' is used as a general term to mean any of the various small-size pyrotechnic or explosive devices. The squib articles contain small quantities of black powder or pyrotechnic substances.

The pyrotechnic squib is a little pipe, or hollow cylinder of paper, filled with powder or combustible matter, to be thrown into the air while burning, so as to burst there with a crack.

As an explosive device squib is used as an initiator to start or deflagration of low explosives. It is used for igniting black powder or pellet powder. A squib is an igniferous device, not a detonator.

In electric squibs, functioning is accomplished by applying an electric current.

Also see: *Fireworks* and *Pyrotechnics*.

Squib, Electric – Electric Squib: In an Electric *Squib*, functioning is accomplished by applying an electric current. Electric Squib means a small tube or block containing a small quantity of ignition compound in contact with a wire bridge. Electric Squibs are constructed very much like electric detonators except that they emit a high temperature flame jet when activated rather than detonating.
Squibs have limited use in commercial blasting and are becoming very difficult to obtain.

Synonym: *Electric Ignitor.*

Also see: *Squib.*

Squib Switch – An electric switch operated by a squib or pressure cartridge.

Synonym: *Explosive Switch.*

Squashed-Head Shell – High Explosive Plastic Shell.

Stability – The ability of an explosive material to retain its original chemical and physical properties without degradation when exposed to specific environmental conditions over a particular period of time in storage. Stability is the resistance of the meta-stable state to heat.

Stability Test – Stability Test means an accelerated test to determine the suitability of an explosive material for long-term storage under a variety of environmental conditions.

Stabilizer – A material that reduces explosive sensitivity.

Synonym: *Moderator.*

Star – Pyrotechnic materials that are discharged from shells, candles, etc., and burn as a single light while in the air. Usually used as signal.

Static Electricity – Static Electricity means the electric charge at rest on a person or object. Electricity energy is stored on a person or object in a manner similar to that of a capacitor. It is most often produced by the contact and separation of dissimilar insulating materials. Static electricity may be discharged into electrical initiators, thereby detonating them.

Steady State Velocity – Steady State Velocity implies the characteristic velocity at which a specific velocity, under specific conditions, in a given charge diameter, will detonate.

Stemming – The insertion of clay or other inert incombustible material into the end of a borehole that resists the pressure of the explosive when the explosive is fired.

Storage – The safekeeping of explosives, usually in specially designed structures named magazines.

Storage Magazine – A structure designed and used for the long-term storage of explosives or ammunition.

Compare: *Storehouse.*

Storage Magazine, In-process – See *In-process Storage Magazine.*

Storehouse – A building other than a magazine for storage of certain types of explosives.

Also see: *Storage Magazine.*

Straight Dynamite – See *Dynamite, Straight.*

Straight Gelatin – Contains nitroglycerin, nitroglycol, or similar liquid sensitizers and sodium nitrate, mixed with nitrocellulose. High detonation velocity. Excellent water resistance.

Straight Gelatin Dynamite – See *Dynamite, Straight Gelatin.*

Stray Current – A flow of electricity outside its normal conductor. Current flowing outside an insulated conductor system. Stray current is a result of defective insulation. Galvanic action of two dissimilar metals, in contact or connected by a conductor, may cause stray current. It may come from electrical equipment, electrified fences, electric railways, or similar items.

Flow is facilitated by conductive paths such as pipelines and wet ground or other wet materials.

Streaming Velocity – See *Velocity, Streaming.*

Strength, Bulk – See *Bulk Strength.*

Strength, Bulk, Absolute – See *Absolute Bulk Strength.*

Strength, Bulk, Relative – See *Relative Bulk Strength.*

Strength, Cartridge – See *Cartridge Strength.*

Strength, Explosive – Explosive Strength: The amount of energy released by an explosive upon detonation, which is an indication of the capacity of the explosive to do work. It is a property, though not a well-defined one, of an explosive described in various terms such as cartridge- or weight strength, shock or bubble energy, crater strength, ballistic mortar strength, etc. (Typically the standard to determine explosive energy is a comparison to the energy of an equal weight of standard ANFO.) It can be the explosive strength of unit weight (or volume) of a high explosive when compared with that of *blasting gelatin* in a ballistic mortar. It is sometimes designated in percentage of nitroglycerine (%NG). This latter designation is not a true measure of its strength.

Also see: *Explosive Energy.*

Strength, Volume – See *Volume Strength.*

Strength, Weight – See *Weight Strength.*

Strength, Weight, Absolute – See *Absolute Weight Strength.*

Strength, Weight, Relative – See *Relative Weight Strength.*

Strike Anywhere Match – See *Match, Strike Anywhere.*

Striker – A part of the firing mechanism of a gun, mine, mortar, etc. which hits the primer, hammer or firing pin of a gun.

Striking Fire – The expression 'Striking Fire' means the act of accidentally causing an explosion while working with pyrotechnical compositions. This is usually caused by steel tools being struck together in the presence of flammable compounds. Sometimes brass and wooden parts may produce enough friction to cause such fire. It is another source of hazard against which the pyrotechnist should be on constant guard.

Strip Mining – See *Opencast Mining.*

Styphnic Acid – Styphnic acid, $C_6H\ (OH)_2\ (HO_2)_3$ is obtained in yellow crystalline form. It melts at 179-180°C, and is slightly soluble in water. It is used in explosives as a priming agent.

Synonym: 2,4,6-trinitroresorcinol.

Styphnate, Lead – See *Styphnate, Lead.*

Sub-machine Gun – See *Gun, Sub-machine*.

Substantial Dividing Wall – Substantial dividing wall is an interior wall designed to prevent the propagation of an accidental detonation on one side of a wall to explosives on the other side.

Successive Explosion – Successive explosions may occur in the case of *rarefied explosion* in confined space. *Rarefied explosions* arise mainly from the ignition of admixtures of flammable gas/vapor, or extremely fine liquid or solid particles with air. A rarefied explosion in a confined space may not always result in complete combustion, and the confined space thereafter may not be safe contrary to common belief. Successive explosions are common with a *dust explosion.* The first dust explosion agitates dust lying on floors, ledges, etc., to form another dust cloud. The resultant dust cloud is ignited by the flame from the first explosion, produces the second explosion - usually of greater intensity. In case of a vessel containing a gas or vapor, it can suffer successive explosions without addition of fresh gas or vapor, if the first explosion occurs with a gas concentration close to the upper explosion limit. The gas or vapor may not be necessarily completely burnt in the first explosion; what remains can often be in concentration above the lower explosive limit, i.e. within the explosive range. Sometimes called a secondary explosion.

Sulfur – Sulfur is an element. It is usually kept in sticks, or in powder form, known as flowers of sulfur. It ignites readily and may, in burning, create inflammable vapors, or, in the form of dust, form an explosive mixture with air. It is dangerous when in contact with oxygen carriers, such as chlorates and nitrates, friction being sufficient to ignite such mixtures. It is an ingredient of black powder. Sulfur may ignite spontaneously in contact with lampblack or charcoal. Gives off noxious fumes when burning, and owing to the low temperature at which it melts, may spread a fire by flowing. Sulfur is used in the manufacture of gunpowder/black powder, sulphuric acid, fertilizer, fireworks, rubber, and as an insecticide.

Synonym: Brimstone.

Sulfur, Flowers of – See *Flowers of Sulfur.*

Super Quick Fuse – A fuse that functions at the quickest possible time upon impact of the missile with the target.

Synonym: Instantaneous Fuse.

Supersensitive Fuse – Supersensitive Fuse means a fuse that sets off a projectile when it strikes even a very light target, such as an airplane wing.

Supplemental Charge – Filler, usually TNT, used in deep cavitied projectiles to fill void between ordinary fuse and booster combination and bursting charge.

Supplementary Explosives Charge – Small removable explosive charges in the cavity of a projectile between the fuse and the main bursting charge.

Support Zone – Synonym of *Cold Zone*.

Surface Magazine – See *Magazine, Surface.*

Surface Flare – See *Flare, Surface.*

Sure-Fire Current – The term 'Sure-Fire Current' means the minimum current that should necessarily be applied to a bridgewire circuit to reliably ignite the prime material without regard to the time of operation.

Sustainer – The term as applied in rocketry is a slow burning motor to produce a continued thrust.

Sustainer Grain – A propellant or pyrotechnic grain used in a pressure cartridge or igniter in order to sustain burning.

Switch, Blasting – See *Blasting Switch*.

Switch, Explosive – See *Explosive Switch.*

Switch, Squib – Squib Switch: An electric switch operated by a squib or pressure cartridge.

Sympathetic Detonation – The detonation or explosion of an explosive material by the detonation of another charge in the vicinity without actual contact. In other words, a sympathetic detonation may termed as the initiation of an explosive charge without a priming device as a result of receiving an impulse from another explosion in the neighborhood through air, earth, or water. It can be said that the second explosion, known as sympathetic one, is initiated by influence.

The distance through which transmission of detonation may occur depends on many factors, such as mass, physical and chemical characteristics, detonation properties of donor charge, existence and characteristics of acceptor charge confinement, existence and characteristics of the medium between the charge, etc. Other relevant factors being constant, the possibility of the propagation of detonation is determined mostly by both the sensitivity of the acceptor charge and by the initiating strength of the donor charge. A possible factor that contributes to sympathetic detonation also includes the initiation of acceptor charge by the passage of shock wave from one mass to the other, by flying fragments and the like.

Synonyms: *Sympathetic Explosion*, *Sympathetic Ignition*, and *Sympathetic Propagation.*

Sympathetic Explosion – Synonym of *Sympathetic Detonation.*

Sympathetic Ignition – Synonym of *Sympathetic Detonation.*

Sympathetic Propagation – Synonym of *Sympathetic Detonation.*

T4 – Synonym of *Cyclonite, Hexogen*, or *R.D.X.*

Taliani Test – A heat stability test for propellants and high explosives.

Tamping – The act of compacting the explosive charge or the stemming in a blast hole, with the aid of a tamping stick.

Tamping Bag – A bag that is used in boreholes to compact the explosive material charge. Tamping bags are cylindrical in shape and contain stemming material.

Tamping Pole – A plastic or wooden pole used to compact explosive charges or stemming.

TATB – Acronym for *Triaminotrinitrobenzene*.

TBI – Acronym for *Through-Bulkhead Ignition.*

Terrorist Bomb – Terrorist bombs are usually custom-made, developed to any number of designs, use a wide range of explosives of varying levels of sensitivity and power, and are used in very many ways. That is why terrorist bombs are generally referred to as improvised explosive devices or IEDs.

Compare: Military Bomb.

Synonym: Improvised Explosive Device (IED).

Tertiary Explosive – The tertiary explosive is a sub-class of high explosives. It is the least sensitive class of explosives when classification is based on the sensitivity to initiation, the primary explosive being the most sensitive explosive. Examples of tertiary explosives are *Ammonium Nitrate - Fuel Oil* (*ANFO*) and *Slurries.*

Tertiary explosives are safe, and easy to make and use. ANFO, for example, may be mixed up with a shovel on the floor or in a cement mixer. But they are not so easy to fire as primary or secondary explosives. A tertiary explosive being the least sensitive high explosive requires an intense shock to initiate. Initiation of tertiary explosives is possible with some high strength detonators, but these are not normally used. Usually, a *primer* or *booster* is used for its initiation with a normal detonator. Tertiary explosives are used as a main charge in boreholes.

Tertiary High Explosive – Tertiary High Explosive is usually shortened to *Tertiary Explosive.*

Test, Abel Heat – See *Abel Heat Test.*

Test Blasting Cap # 8 – See *IME # 8 Test Detonator.*

Test, Bomb Drop – See *Bomb Drop Test.*

Test, Clearing – See *Clearing Test.*

Test, ADC – See *ADC Test.*

Test, Fall Hammer – See *Fall Hammer Test.*

Test, Hess – See *Hess Test.*

Test, Jolt and Jumble – See *Jolt and Jumble Test.*

Test, Kast – See *Kast Test.*

Test, Stability – See *Stability Test.*

Test, Talini – See *Talini Test.*

Test, Torpedo – See *Torpedo Test.*

Test, Torpedo Friction – See *Torpedo Friction Test.*

Test, Trauzl – See *Trauzl Test.*

Tetracine – Synonym of the chemical compound *Guanylnitrosoamino-guaanyltetrazene.*

Tetralite – Synonym of *Tetryl.*

Tetranitroaniline – Tetranitroaniline (TNA) is a powerful and sensitive high explosive, similar to *tetryl.* It is used as a booster for high explosive shells and in primer and detonating compositions. It deteriorates in the presence of moisture.

Tertranitrodiglycerin – It is a less sensitive high explosive. It resembles nitroglycerin in many respect. Due to its own low-freezing point, tetranitrodiglycerin is used as a component of low-freezing dynamites.

Tetranitromethane – It is the tetranitro compound of well-known methane gas. Tetranitromethane possesses a large excess of oxygen and hence it can form very powerful explosives when mixed with other nitro compound high explosives that are somewhat oxygen deficient. It is used in blasting explosives and in detonating compositions.

Tetranitronaphthalene – It is a high explosive. It is equal to but somewhat less sensitive to impact than TNT. It is much used for bursting charges.

1,3,5,7-tetranitroperhydro-1,3,5,7-tetrazocine – IUPAC name for *HMX.*

Tetril – Synonym of *Tetryl.*

Tetryl – Chemical name of Tetryl is 2,4,6-Trinitrophenylmethylnitramine ($C_7H_5N_5O_8$), $(NO_2)_4$ C_6H_2 - N - CH_3. It is obtained, as a fine yellow crystalline solid material, by nitration of dimethylaniline. m.p. 129° - 130°C. When tetryl is heated, it first melts, then decomposes and explodes.

This chemical compound is a powerful explosive. When tetryl is heated, it first melts, then decomposes and explodes. It is fairly stable, but quite sensitive to percussion and it can be initiated from flame, friction, shock, or sparks. It is more sensitive than *Explosive D* or *TNT.*

It can be compressed into pellets for use as a booster explosive where stable explosives need more than simply an initiator to cause them to detonate. It is considered the standard booster charge for high explosive shells. It is used in reinforced detonators, and as the explosive filler in some types of projectiles.

Due to some environmental hazards, Tetryl is no longer manufactured in the United States.

Synonyms: Nitramine, Tetralite, Tetril, and Trinitrophenylmethylnitramine.

Tetrytol – A military high explosive comprised of a mixture of *Tetryl* and *TNT* usually in the proportion of 75% and 25% respectively. The explosive force of tetrytol is approximately the same as that of TNT. It may be initiated by a blasting cap. Tetrytol is usually loaded by casting. This binary explosive is used in artillery shells, as a bursting charge for mines, and a demolition explosive.

Theft-resistant – Construction designed to deter illegal entry into facilities used for the storage of explosive materials.

Thermal Detonator – A thermal detonator is a compact yet powerful explosive device contained is a small, silvery sphere. Once activated, an internal fusion reaction starts within the sphere, which eventually grows into a deadly explosion.

Thermal Explosion – The explosion that is caused by a *thermal detonator*.

Thermite – A high temperature producing mixture comprised of iron oxide and powdered aluminum, which when burning produces a very high temperature of about 2500°C. It cannot be extinguished, either by water or by smothering. Thermite forms an incendiary composition consisting of 1.0 part of granular aluminum and 2.75 parts ferrosoferric oxide (black iron oxide). It is sometimes used for welding and also as a filling for magnesium incendiary bombs.

Thermonuclear Bomb – Synonym of Hydrogen Bomb.

Also see: *Nuclear Weapon*.

Thermos Bomb – An *anti-personnel bomb* used during the World War II by the Italian Air Force. The bomb was so named because of its resemblance to a popular brand of vacuum flask, Thermos. The bomb could be fitted with a highly sensitive fuse that would detonate if an attempt was made to move it, and could kill someone in the open up to about 35 metres (115 feet) away.

Through-Bulkhead Ignition – The expression 'Through-Bulkhead Ignition' implies a means of transferring a detonation from one side of a bulkhead to the other without destroying the integrity of the bulkhead seal.

Acronym: *TBI*.

Time, Application – See *Application Time*.

Time Bomb – A bomb having a device to set to explode at a predetermined time.

Time bombs are generally improvised explosive devices consisting of batteries, detonators, explosive charge, and timer. Earlier types contained safety fuse.

Synonym: Timer Delay.

Time, Bursting – See *Bursting Time.*

Time, Dwell – Dwell Time: In press loading powders into cavities, the interval of time that the powder is held at the full loading pressure.

Time, Excitation – Excitation Time: In firing electric detonators, the interval between the application of the current and the firing of the fuse head.

Time Fuse – A fuse made to burn for a given time, especially to explode a bomb.

Time Fuse – A fuse designed to burn for a certain period of time before producing an explosion. This is accomplished either by varying its length or by the changing the character of its composition.

Time, Lag – See *Lag Time.*

Time, Reaction – See *Reaction Time.*

Timer Delay – Synonym of *Time bomb.*

TNA – Acronym for *Tetranitroaniline*.

TNB – Acronym for *Trinitrobenzene*.

TNP – Acronym for *Trinitrophenol.*

Synonym: *Picric Acid.*

TNT – Acronym for *Trinitrotoluene*, or *Trinitrotoluol.*

TNT Equivalent – The term 'TNT Equivalent' is used as a measure of the blast effects from the explosion of a given quantity of material expressed in terms of the weight of TNT that would produce the same blast effects when detonated.

TNX – Acronym for *Trinitroxylene.*

Tommy Gun – A synonym of the Thompson submachine gun. Invented in 1920s, the easily portable automatic weapon fired 450 to 600 cartridges per minute.

Also See: *Gun.*

Tonite - A compressed explosive compounded of guncotton and barium nitrate. Tonite is used for blasting, for distress signal work, and for submarine demolition.

Torches – Paper tubes, 1.25 to 2.50 centimeters (1/2 to 1 inch) in diameter, charged with pyrotechnical compounds, usually colored and used in parades, etc.

Toroid Induction – It is an electric detonator to which a ferrite ring/toroid is attached. The ferrite ring/toroid serves as a transformer to convert incoming high frequency AC energy into current to energize a conventional bridgewire electric initiator. A specialized high frequency power source is required for this device. This type of detonator is available in short period as well as long period delays with a selection of leg wire lengths. This type of initiator provides increased protection against extraneous current sources, ease of hook up, and a degree of protection against current leakage.

Torpedo Explosive No. 1 – Commercial term for *Blasting Gelatin.*

Torpedo, Railway – The term 'Railway Torpedo' is a synonym of *Railway Fog Signal***.**

Torpex Booster – A *booster* consisting of a mixture of *RDX, TNT*, and *Aluminum.*

Torpedo – A self-propelled and usually directable, underwater missile designed to contain an explosive charge launched from a tube located on the deck or inside the hull of a warship. Torpedoes are chiefly used in war against ships transporting troops and supplies.

The first stationary torpedo was invented by Captain David Bushnell, who tried it out in New York Harbor in 1776. It was named after the torpedo fish. The first invented torpedo comprised of an explosive charge fixed to an enemy hull and was set-off by a clockwork fuse. The early stationary torpedoes exploded against vessels. Those were later classified as mines. The English engineer Robert Whitehead developed the first automatic torpedo in 1868. The Russian navy became the first user of automatic torpedoes when they sank a Turkish ship during a war in 1878. The German Navy became the first to fire an automotive torpedo in 1914 when a German U-boat unsuccessfully attacked a British battleship.

Torpedo, with Bursting Charge – These are devices containing means of propulsion and charge of secondary detonating explosive.

Torpedo Friction Test – A test of explosives to determine sensitiveness to impact and friction.

Total Content – Such a substantial proportion, assuming simultaneous explosion of the whole of the explosives. Used to assess the practical hazard.

Synonym: Entire Load.

Tracer – Element of a type of ammunition (called tracer ammunition) containing a chemical composition that burns visibly in flight. Tracer is used for observation and adjustment of fire, for incendiary purposes, and for signaling.

Tracer for Ammunition – Devices containing pyrotechnic composition, designed to reveal the trajectory of a projectile, which do not contain exposed pyrotechnic material.

Tracer Fuse – A 'Tracer Fuse' means a device attached to any projectile and containing a slow-burning composition.

Train, Burning – See *Burning Train.*

Train Component, Explosive – A device containing an explosive and designed to transmit the detonation within an explosive train.

Train, Explosive – Explosive Train: It is a sequence of events that begin with relatively high level of sensitivity (i.e., low levels of energy) that causes a chain reaction to initiate the final *main charge*. It can be a low or high explosives train.

Explosive train is, in effect, a series of explosive elements consisting of the initiator, the booster, and the main (burster) charge. However, it is often compounded by the addition of intermediate charges and time delays. The elements are arranged in order of decreasing sensitivity. The explosive train accomplishes the controlled augmentation of a small impulse into an impulse of suitable energy to cause the main charge to function. In other words, the process in which the *primary explosive* sets off the *booster*, which in turn sets off the insensitive *main charge*, may be termed as the explosive train.

The basic high explosive train consists of the detonator, booster, and bursting charge. High explosives trains may also be either two-step ones, such as detonator and dynamite. A typical example of three-step high explosive train is detonator, booster and ANFO.

A fuze explosive train may consist of a primer, a detonator, a delay, a relay, a lead and booster charge, one or more of which may be either combined with another element or omitted.

Low explosive trains are something like a bullet - primer and a propellant charge. The sequence of events in small arms ammunition are: (a) Flame/Heat ignites the propellant charge, (b) Production of Gases, and (c) Gases drive the bullet through the bore of the weapon

The gun propellant requires the use of another type of explosive train. The small flame produced by the initiating charge is insufficient to ignite the propellant grains thoroughly to produce an efficient burning rate of the entire charge. A propelling charge explosive train consists of a primer, igniter or igniting charge, usually a mixture of lead styphnate and nitrocellulose, and finally the propellant charge.

Synonym: *Firing Train*; *Initiation Sequence.*

Train, Firing – Synonym of *Explosive Train.*

Train, High Explosive – See *High Explosive Train.*

Train, Igniter – See *Igniter Train.*

Train, Powder – See *Powder Train.*

Trajectory – Path of bomb, missile or projectile in flight.

Transmission of Detonation – Synonym of *Sympathetic Detonation.*

Transmission of Explosion – Synonym of *Sympathetic Detonation.*

Trauzl Test – Method of determining relative energy available from an explosive material by measurement of the volume expansion of a lead block test.

Triaminotrinitrobenzene – 1,3,5-triamino-2,4,6-trinitrobenzene (TABT) is a high explosive. It is extremely insensitive. It is virtually invulnerable to significant energy release in fires, and explosions or to deliberate attack with small arms fire. Initiating a TATB detonation is not easy. Detonation is still fast but by no means instantaneous. The resulting shock wave propagates differently from that of sensitive explosives, and the molecules that detonation produces are different.

Triangle of Combustion – Synonym of *Fire Triangle.*

Triangle of Explosion – See *Explosion Triangle.*

Triangle of Fire – See *Fire Triangle.*

Trimethylene Trinitramine – See *Cyclonite* or *Hexogen*, the chemical name of Cyclonite and Hexogen.

Trimonite – A binary high explosive comprised of a mixture of picric acid and mononitronaphthalene. Trimonite is used as a substitute for trinitrotoluene as a bursting charge.

Trinitroanisole – Trinitroanisole is a high explosive compound that resembles picric acid in its high explosive properties, but does not attack metals provided it is protected form moisture. It has a lower melting point than picric acid, which is an advantage in shell loading. It is used as a booster charge.

Trinitrobenzene – Trinitrobenzene (TNB) is the simplest of the aromatic explosive compounds, and structurally the basis for all of the others in this family.

NO_2 NO_2

NO_2

TNB

It is considered a powerful high explosive and has more shattering power than TNT. It is less sensitive to impact and probably more toxic than TNT. However, it is not often used because it is difficult to produce.

Trinitrochlorbenzene – Trinitrochlorbenzene, [$C_6H_2Cl(NO_3)$] is a typical organic nitrate. It produces phosgene, which is a poison gas, upon detonation. Because of this property it is an interesting explosive for military purposes, and hence it was often used in World War I.

Trinitrocresol – Trinitrocresol is not as powerful a high explosive as TNT or picric acid. It has been used as a bursting charge and in combination with other high explosives.

Synonym: *Cresolite.*

Trinitroglycerin – Synonym of Nitroglycerin.

Trinitrophenol – See *Picric Acid,* the synonym of *Trinitrophenol.*

Trinitrophenylmethylnitramine – See *Tetryl*, the synonym of this chemical compound.

Trinitrophenylnitramine Ethyl Nitrate – This chemical compound is a high explosive. It is soluble in nitroglycerin. It explodes when heated. Its high explosive sensitivity to impact and friction are about the same as that of tetryl, but its shattering power is much greater. Trinitrophenylnitramine ethyl nitrate is used as a base-charge in detonators.

Synonym: *Petryl.*

Trinitroresorcinol – See *Styphnic Acid,* the synonym of this explosive compound, better known as 2, 4, 6-trinitroresorcinol.

Trinitrotoluene – 2,4,6-trinitrotoluene, better known as TNT, is obtained by the nitration of toluol (toluene), in the form of yellow prismatic crystals. Its melting point is 81°C. It is soluble in alcohol and ether but not in water. TNT is an example of mono-substituted *TNB.*

TNT was discovered in 1863 as a dye agent. It was not used as an explosive until 1904. It has been the most widely used military explosive since World War I. It had become the standard military explosive by World War II. The power of other explosives is frequently expressed as an equivalent amount of TNT.

TNT is extremely moisture resistant and is not likely to be detonated by physical shock. It is very stable and can be stored for long periods. The advantages of TNT include safety in handling, fairly high explosive power, good chemical and thermal stability, compatibility with other explosives, and moderate toxicity. Due to its low melting point, it can be cast easily by melting the material and then it can be poured into shells.

TNT is commonly used as a filling for shells. It is widely used as explosive filler in projectiles and by engineers. TNT is a military explosive compound used industrially as a sensitizer for slurries and as ingredient in pentolite and composition B. It is also used as a free running pelletized powder.

Acronym: TNT.

Synonyms: *Trinitrotoluol*, *Triton*, and *Trotyl.*

Trinitrotoluol – Synonym of *Trinitrotoluene.*

Trinitroxylene – Trinitroxylene (TNX) is a high explosive compound. It is an example of poly-substituted TNB. This high explosive is not very powerful when used alone. However, the addition of picric acid or other nitro-type of high explosive serves to lower its melting point and reinforce its explosive power. It is used in mixtures with ammonium nitrate; in detonating compositions, and mixed with other high explosives, as a bursting charge.

Triple-Base Propellant – Propellant whose principal active ingredients are nitrocellulose, nitroglycerin and nitroguanidine.

Also see: *Propellant.*

Triple Point – Intersection of the original shock wave, the reflected shock wave and the Mach wave.

Triton – Synonym of *Trinitrotoluene (TNT).*

Tritonal – Tritonal is a binary high explosive composed of 80% of TNT and 20% of Aluminum. It is a silvery solid that is used as filler in bombs and shells.

Trotyl – Synonym of *Trinitrotoluene (TNT).*

Trunkline – A detonating cord line on the ground surface used to connect the downlines or other detonating cord lines in a blast pattern. Generally trunk line runs along each row of blast holes.

Tube, Blast – See *Blast Tube.*

Tube, Priming – Priming Tube: A small pipe, filled with a combustible composition for firing cannon.

Tube, Quincke – See *Quincke Tube.*

Tube, Safety – Safety Tube: Safety tube means a tube that is designed and used to prevent explosion or to control delivery of gases by an automatic valvular connection with the outer air.

Tube, Shock – Shock Tube: Shock tube is a small diameter plastic laminate tube coated with a very thin layer of reactive material, such as HMX and Aluminum Powder, at 1 kg per 67,100 meters (1 pound per 100,000 feet) of tube. When initiated, shock tube transmits a low energy signal or propagation wave, at approximately 2000 meters (6,500 feet) per second from one point to another. This shock wave is similar to a dust explosion and will propagate through most sharp bends, knots and kinks in the tube. The detonation is sustained by such a small quantity of reactive material; the outer surface of the tube remains intact during and after functioning. It is very insensitive to initiation by ordinary heat or impact and requires an intense high impulse shock to be energized. Shock tube can be initiated by small shot shell-type primers, detonating cord or detonators.

Shock tube is used to initiate boosters, detonating cord, other lengths of shock tubes and other explosives. This type of initiation is common to the construction industry.

Tube, Shock, Detonator – See *Shock Tube Detonator.*

Tube, Shock, System – See *Shock Tube System.*

Tubular Grain – A solid propellant grain in the form of a tube.

Tubular Primer – Synonym of *Primer for Propelling Charge.*

Tun Dish – A large unstirred cylindrical vessel used for slow processes of steeping or steaming.

Two-component Explosive – See *Binary Explosive.*

Udex Process – A proprietary petroleum refining process used to extract aromatic hydrocarbons from a complex hydrocarbon mixture. Glycol-water solvent is used in this process.

UEL – Acronym for *Upper Explosion Limit.*

UFL – Acronym for *Upper Flammability Limit.*

Ullage – The empty volume of a propellant tank that is not occupied by fuel or oxidizer.

Ultra Fining – A proprietary fixed-bed catalytic hydrogenation process used in petroleum refineries to desulfurize and upgrade naphtha distillates.

Ultraforming – A proprietary catalytic reforming process used to increase the octane ratings of petroleum naphthas.

Umbrite – An explosive containing 49 parts of *nitroguanidine*, 38 parts *ammonium nitrate*, and 13 parts of silicon; slightly hygroscopic, and retaining its explosive power even when moist.

Unabomber – A terrorist who periodically mailed package-bombs to victims at universities and airlines, as well as other organizations in the USA. Dubbed by the Federal Bureau of Investigation as 'unabomber', which stands for universities and airlines bomber.

Uncharged Spices – A chemical entity with no net electric charge.

Synonym: Neutral Species.

Undercutting – 1. In coal mining, the production of a free face by cutting out mechanically the lowest few inches of a seam.

2. In distillation, the technique of taking the products coming off the distillation tower at a temperature below the desired ultimate boiling point range to prevent contaminating the products with the compound that would distill just beyond the ultimate boiling point range.

Unconfined Detonation Velocity – The detonation velocity of an explosive material not confined by a borehole or other confining medium. A charge fired in the open is an example of unconfined detonation.

Unconfined Explosion – An unconfined explosion is an explosion occurring in the open air where the (atmospheric) pressure is constant. Such explosion does not occur in a substantial container or in a blast hole. *Open air explosion* is an example of unconfined explosion.

Unconfined Vapor Cloud Explosion – The term 'Unconfined Vapor Cloud Explosion' is an imprecise one originally used as a definition for 'vapor cloud explosion', but not commonly used now. A 'vapor cloud explosion' is sometimes termed 'unconfined vapor cloud explosions'. It is also described as 'open air explosion'.

Acronym: UVCE.

Synonyms: *Open Air Explosion, Vapor Cloud Explosion*, and *VCE.*

Also see: *Vapor Cloud Explosion.*

Uncontrolled Explosion – Accidental physical explosions, or rarefied chemical explosions, or nuclear explosions (either fission or fusion or both) are examples of uncontrolled explosions.

Underground Magazine – A specially designed and constructed structure for the storage of explosive materials underground.

Under Sonic Propagation – Synonym of *Deflagration.*

Undesired Explosion – Physical explosions or accidental chemical explosions are examples of undesired explosions.

Unexploded Ordnance – Explosive weapons (bombs, shells, grenades, etc.) which have been primed, fused, armed, or otherwise prepared for action, and which has been fired, dropped, launched, projected, or placed in such a manner as to constitute a hazard to operations, installations, personnel, or material, and remains unexploded either by malfunction or design or for any other cause. Unexploded ordnances still pose a risk of detonation, even decades after the battles in which they were used.

Acronym: UXO.

Unfired Pressure Vessel – A pressure vessel that is not in direct contact with a heating flame.

Unisol Process – A proprietary solvent extraction process used in petroleum refineries to extract mercaptan sulfur and certain nitrogen compounds from sour gasoline or distillate.

Unit, Auxiliary Power – Auxiliary Power Unit: It is a propellant-powered device used to generate electric or fluid power.

Acronym: *APU.*

Unit, Explosive – See *Explosive Unit.*

Unit Load – A load comprising a number of suitable outer packages of compatible explosive substances or articles combined for ease of transport, designed to be carried, stored, and handled as a separate unit and able to withstand the conditions associated with the appropriate modes of transport.

Upper Explosion Limit – Upper Explosive Limit (UEL) is the maximum concentration of vapor in air, which forms an explosive or combustible mixture.

Upper Flammability Limit – Upper Flammable Limit (UFL) is the maximum concentration of vapor in air, which forms an explosive or combustible mixture.

Uranium Carbide – One of the carbides of uranium, such as uranium monocarbide. Uranium carbide is used chiefly as a nuclear fuel.

Uranium Hydride – Uranium hydride, UH_3 is a highly toxic, gray to black powder that ignites spontaneously in air. It conducts electricity. It is used for making powdered uranium metal, for hydrogen-isotope separation, and as a reducing agent.

Uranium Nitrate – See Uranyl Nitrate

Uranyl Nitrate – Uranium Nitrate, $UO_2(NO_3)_2.6H_2O$. It is an unstable yellow crystalline salt, which melts at 60°C and boils at 118°C. It has toxic and explosive characteristics. It is soluble in water, alcohol, and ether. Yellow salt is used in photography, in medicine, and for uranium extraction and uranium glaze.

Synonym: Uranium Nitrate, and Yellow Salt.

NH

Urea Nitrate – Urea Nitrate, $O{=}C<$

$NH_2{\cdot}HNO_3$,

$CO(NH_2)_2.HNO_3$ or $CH_5N_3O_4$,is a colorless crystalline compound. It has fire-hazard and explosive characteristics. Its stability is poor; decomposes at 152°C. It is soluble in alcohol, slightly soluble in water. It is used in explosives and to make urethane.

Urea Peroxide – Urea Peroxide, $CO(NH_2)_2.H_2O_2$ is an unstable white powder chemical compound which decomposes at 75-85°C or by moisture. It has fire-hazard characteristics.

Use of Explosives – Explosives have found applications both for the purpose of strengthening man's arm in war and for many peaceful uses. Explosives were first used for military purposes. The destructive effects of explosives are much more spectacular than their peaceful uses.

Explosives are of immense value in many peaceful pursuits. Explosives are the basis for many of mankind's most constructive efforts, as explosives may be treated as an engineering tool. Three tunnels were constructed by the use of explosives and even today stand out as benchmarks in the history of explosives: Mount Cenis, a 13-kilometer railway tunnel driven through the Alps between France and Italy (1857-71); the 6.4-kilometer Hoosac Railway Project; and the Sutromine Development Tunnel in Nevada (1864-74).

Peaceful use of explosives is not limited to mining, quarrying, and engineering enterprises, but also in making fireworks, signal lights, and rockets. Explosives are used to project lifelines to ships in distress off storm-beaten shores and to break up ice jams, to extinguish oil well fires, inflate automobile air bags and destroy hazardous wastes. When pile drivers are not available, exploding dynamite on an iron plate placed on top of the piles can do their work. Farmers find explosives useful for breaking up boulders, blowing out stumps, felling trees, and loosening soil. Blind rivets are needed when space limitations make conventional rivets impractical. Explosive riveting is an engineering practice.

Explosives are sometimes used to bond various metals to each other. The very fine industrial-type diamonds used for grinding and polishing are produced by the carefully controlled action of explosives on carbon.

UXO – Acronym of *Unexploded Ordnance.*

Vacuum Bomb – The *fuel-air explosive* is popularly known in Russia as 'vacuum bomb'. The United States first developed and used fuel-air explosives in Vietnam War. Soon the Russian scientists developed their own FAE weapons.

FAE uses the chemicals that detonate when mixed with ambient air, which is required to maintain the oxygen balance for detonation. Ethylene oxide and propylene oxide are common FAE fuels.

A typical fuel-air explosive device consists of a container of fuel and two separate explosive charges. After the munitions are fired or dropped, the first explosive charge bursts open the container at a predetermined height and disperses the fuel in a cloud that mixes with atmospheric oxygen. The cloud of fuel flows around objects and into structures. The second charge then detonates the cloud, creating a massive blast wave that destroys unreinforced buildings and equipment and kills and injures personnel. The pressure wave is responsible for killings; still more important is the subsequent rarefaction [vacuum], which ruptures the lungs. Even if the fuel deflagrates instead of detonates, victims will be severely burned and will probably also inhale the burning fuel. Since the most common FAE fuels are highly toxic, an undetonated FAE is also lethal to personnel caught within the cloud as most chemical agents.

Synonym: Fuel-air Explosive

Valve, Safety – *Safety Valve*: 1. A type of pressure relief valve intended for the release of excessive vapor pressure such as for a steam boiler or hydraulic system. Such an automatic escape or relief valve is held shut by an arrangement exerting a definite pressure so that the valve lifts, and the steam, water, or other contents escape as and when the pressure exceeds a predetermined amount. 2. A similar automatic valve opening inward in order to admit air to a tank or vessel in which the pressure has become less than that of the atmosphere and prevents collapse.
[LCS1]
Vapor Cloud Explosion – A chemical explosion due to an ignition in an unconfined cloud, made up of a mixture of a flammable vapor or gas with air.

The discovery of vapor cloud explosion (VCE) is relatively a recent one. Strehlow in 1972 was able to demonstrate that VCEs could occur and that they represented a very serious hazard. The Fixborough (U.K) explosion of 1974 was recognized as being an example of a VCE. No clear example of a VCE can be demonstrated prior to 1943. It is mentionable that the phenomenon of confined explosions has been recognized for at least two centuries.

The durations of VCE blast waves may be expected to be longer than those associated with dense explosives. A stick of dynamite takes about 10^{-5} seconds to detonate, whereas a vapor cloud of, say, 200 meters diameter will take about 1 second for a flame front traveling at about 0.5 times sonic velocity to cross it. A vapor cloud may have a diameter of the order of hundreds of meter, whereas only a very large explosive charge would have a diameter exceeding 3 meters. Unlike *dense explosions* VCEs do not have a sharply defined center. Although VCEs may give rise to extensive damage, VCEs are of very low brisance and do not give rise to craters. However, the pattern of damage is not the same as there is no highly brisant center in a VCE.

Synonyms: *Unconfined Vapor Cloud Explosion*, *UVCE*, and *Open Air Explosion.*

Vapor Cloud Explosion, Unconfined – See *Unconfined Vapor Cloud Explosion.*

Vapor Explosion – A chemical explosion due to an ignition of a mixture of a flammable vapor with air.

Vapor, Ghost – See *Ghost Vapor.*

Vapor Pressure – The force exerted when a solid or liquid is in equilibrium with its own vapor, depending on its composition and temperature.

Variable-Time Fuse – See *VT Fuse.*

Vasodilator – Trinitroglycerin, the active ingredient of dynamite is also used medically as a vasodilator. A very useful heart drug is made up of trinitroglycerin. The low concentration of nitroglycerin in medication, and the moist environment in the body, make it impossible for it to explode. The nitroglycerine that is absorbed by the body dilates blood vessels, increasing the blood supply to the heart, and reducing the workload of sending blood around a body. Angina pain is relieved as the heart muscle receives the oxygenated blood it needs for good functioning.

Velocity, Detonation, Confined – Confined Detonation Velocity: The detonation velocity of an explosive or blasting agent in a substantial container or a borehole.

Also see: *Velocity of Detonation (VOD).*

Velocity, Detonation, Unconfined – See *Unconfined Detonation Velocity.*

Velocity, Detonation – Detonation Velocity: See *Velocity of Detonation (VOD).*

Velocity of Detonation – The rate at which the detonation wave travels through a high explosive. The term 'velocity of detonation' relates only to high explosives whereas *velocity of explosion (VOE)* refers to both high and low explosives.

It may be measured confined or unconfined. The unit of the rate of the reaction is meters/second (or feet/second). Velocity of Detonation (VOD) of high explosives shock wave usually varies from 2,000 meters/second (6500 feet/second) to 8,000 meters/second (26,000 feet/second), the speed of sound in air being 331 m/second (1085 feet/second).

Also see: *Velocity of Explosion.*

Velocity of Explosion – The rate at which the combustion/detonation wave travels through the explosive product. It denotes the speed or how fast the chemical reaction occurs or the rate of the reaction. Velocity of Explosion (VOE) refers to both high and low explosives. It is measured in meters/second (or feet/second).

Also see: *Velocity of Detonation.*

Velocity, Steady State – See *Steady State Velocity.*

Velocity, Streaming – Streaming velocity: The velocity of the products of detonation in the direction of travel of the detonation wave.

Veltex No. 448 – An explosive mixture consisting of HMX (octogen) 70%, nitrocellulose (13.15%N) 15%, nitroglycerine 10.7%, triacetin 3.0%, and 2-nitrodiphenylamine 1.3%.

Vessel, Pressure – See *Pressure Vessel.*

Vessel, Pressure, Unfired – See *Unfired Pressure Vessel.*

Vest, Suicide – Suicide Vest: A vest packed with explosives and armed with a detonator, worn by suicide bombers. An explosive belt is usually packed with bolts, nails, screws, and other objects that serve as shrapnel to maximize the number of casualties in the explosion.

Synonyms: Explosive Belt, and Suicide Belt.

Vinyl Benzene – Vinyl benzene, $C_6H_5CH:CH_2$ is a derivative of benzene. It is a colorless, toxic liquid with a strong aroma. It boils at 145°C; is insoluble in water but soluble in alcohol and ether. Vinyl benzene polymerizes rapidly, and can become explosive. It is used to make polymers and copolymers, polystyrene plastics, and rubbers.

Vinyl Chloride – Vinyl Chloride, CH_2: CHCl is a flammable, explosive gas with an ethereal aroma. It boils at $^{-}14$°C. It is soluble in alcohol and ether, slightly soluble in water. Vinyl chloride is an important monomer for polyvinyl chloride and its copolymers. It used in organic synthesis and in adhesives.

Synonym: Chloroethene, and Chloroethylene.

Venturi Loader – See *Jet Loader.*

Very Pistol – A device for firing pyrotechnical cartridges.

VOD – Acronym for *Velocity of Detonation.*

VOE – Acronym for *Velocity of Explosion.*

Volume Strength – See *Cartridge Strength,* or *Bulk Strength.*

VT Fuse – The VT fuse (short for variable-time fuse) is a proximity fuse developed during World War II as a means to improve anti-aircraft gun accuracy. This fuse is designed to detonate an explosive automatically when close enough to the target to destroy it.

W I – Permissible/permitted explosives belonging to the German safety class I.

W II – Permissible/permitted explosives belonging to the German safety class II.

W III – Permissible/permitted explosives belonging to the German safety class III.

Wagon, Fire – See *Fire Wagon.*

Wall, Dividing, Substantial – See *Substantial Dividing Wall.*

Warhead – The article that contains secondary detonating explosives, designed to be fitted to a rocket, guided missile or torpedo. Warheads may be fitted only with a burster or expelling charge.

Warning Signal – A signal, either visual or audible, that is used for warning personnel about the impending explosion in the vicinity of a blast area.

Warning Signal, Lightning – See *Lightning Warning Signal.*

Wasacord – A brand of *Detonating Cord*, which contains about 12 grams of PETN per meter.

Water-activated Contrivance – A contrivance whose functioning is dependent on the physico-chemical reaction with water.

Water Bomb – Water bomb is not an explosive bomb. It is a toy-related item. It is a simple small latex rubber balloon filled with water, or a roughly spherical container made of paper capable of holding water for some time. The user then throws the water-filled balloon or the spherical container made of paper at a desired target and it pops, leaving the target soaked with water. Water bombs are commonly used by children in outdoor play-fights, and by others in carrying out practical jokes.

Water Gel – An explosive material consisting of a mixture of water, oxidizers, fuels, thickeners plus a cross linking agent. This aqueous mixture has a gelatinous consistency. This group of explosives is often referred to as *slurries.*

Ammonium nitrate is generally used as the oxidizer in water gel, which forms the bulk of the mixture. Fuels are used as sensitizers. The fuels may be non-explosive ones such as carbon, sulfur or coarse grade aluminum. A high explosive sensitizer, such as TNT or smokeless powder or monomethylamine nitrate may also be used. Sometimes aluminum powder is added to the mixture to provide further energy when required. Guar gum is added to thicken the solution, which is stabilized by adding potassium dichromate.

It may be an explosive or a blasting agent. That is why some water gel explosives require a primer charge to initiate them. When a high explosive sensitizer is used, it is detonator sensitive, and hence it is an explosive. When a non-explosive fuel is used, it is not normally detonator sensitive, and hence it is a blasting agent.

The ratio of the constituent of water gels may vary, which usually remains confidential to each manufacturer. However, ammonium nitrate is up to 15-20% water, with an added quantity, usually 15-20%, of either calcium nitrate or sodium nitrate. Since the gel base content of a water gel absorbs energy in evaporating its water during the explosive reaction, it is kept to a minimum.

Water gels were developed to overcome the deficiencies of ANFO in wet conditions. Its water resistance is excellent to very good.

Water gels are available as a packaged detonator sensitive explosive or a bulk or packaged blasting agent.

Also see: *Dynamite*, *ANFO*, *Slurry*, *Emulsion*, and *Cast Booster.*

Water Resistance – A qualitative measure of the ability of a blasting agent or an explosive to withstand exposure to water without becoming desensitized or deteriorated.

Wave – The term 'wave' implies a disturbance in a medium or in space which involves the sometimes elastic displacement or vibration of material particles, displacement of electric and magnetic vectors (electromagnetic waves), or a periodic change in some physical quantity, such as temperature, pressure, electric potential, electromagnetic field strength, etc.

Wave, Blast – Blast Wave: A pressure pulse created by an explosion. Blast wave is formed by the rapid expansion of hot gases in the surrounding atmosphere (medium) as a result of an explosion. It is initially a shock wave that subsequently decays into a sound wave.

A large amount of energy is created in a short space of time when a chemical explosion occurs. The energy comes in the form of heat and generally a huge quantity of gas. The explosion reaction occurs so quickly that the gases do not expand immediately, but when they expand they create an enormous 'blast wave.' Hence, it is a wave of high amplitude, a sharply defined wave of increased pressure rapidly propagated through a surrounding medium from a center of detonation or similar disturbance.

Apart from evolution of gases from a chemical explosion, blast waves may be created by other means, such as a physical explosion of a pressurized system when a gas under pressure is released suddenly into the atmosphere. A rapid phase transition, in which the change of state occurs as a result of a significant temperature difference between two or more substances as they come into contact, may produce a blast wave, such as the instantaneous vaporization of water to steam on contact with molten metal.

A blast wave consists of an initial positive pressure phase followed by a negative pressure phase. As and when the pressure pulse formed by a blast wave creates a sharp discontinuity, this is usually termed a shock wave. Blast wave is initially a shock wave that subsequently decays into a sound wave.

Also see: *Shock Wave.*

Wave, Detonating – Detonating Wave: The shock wave produced when a detonator is ignited.

Wave, Detonation – Detonation Wave: A detonation wave is a *shock wave* in a reacting explosive material where the chemical reaction is undergoing in the shock front.

In other words, a detonation wave can be regarded as consisting of an extremely strong shock wave closely followed by an exothermic reaction capable of providing enough energy to sustain the wave. It is far more violent and destructive than a shock wave, due to the presence of a greater amount of energy produced from the reaction, and the chemical combustion is continuously initiated by the adiabatic compression and heating of the gaseous medium behind the shock front.

A detonation wave is the most severe of all the combustion processes. A supersonic combustion consists of a shock front closely followed by a reaction - the two components of a detonation wave.

Wave, Mach – Mach Wave: It means supersonic shock wave.

Wave, Percussion – Percussion Wave: A *shock wave* from an explosion or a blow.

Wave, Secondary – Secondary Wave: A wave created when the air around the blast rushes in to fill the vacuum created during the original explosion.

Wave, Shock – Shock Wave: It is a pressure pulse formed by an explosion in which a sharp discontinuity in pressure is created as the wave travels through a fluid medium at a supersonic velocity. In short, a pressure pulse that propagates at supersonic velocity is termed a shock wave.

A shock wave may also be generated by violent disturbances in a fluid, such as a lightning stroke or a bomb blast.

A detonation of a high explosive generally occurs when a shockwave passes through a block of the high explosive material. In a high explosive, the fuel and oxidizer are chemically bonded, and the shockwave passing through that material at a supersonic rate breaks apart the bonds and re-combines on rearrangement the two materials to produce mostly gasses.

Also see: *Blast Wave.*

Weapon – The instrument used for the defensive and offensive combat. Swords, Grenades, and Rifles are examples of weapons.

Synonym: *Arm.*

Weapon, Cartridge for – Cartridge for Weapon: Fixed (assembled) or semi-fixed (semi-assembled) ammunition designed to be fired from weapons of caliber larger than 19.1 mm. Each cartridge includes all the components necessary to function the cannon once. In the case of cartridges for weapons with inert projectiles, the presence of a tracer can be disregarded for classification purposes provided that the predominant hazard is that of the propelling charge.

See *Cartridge for Weapon.*

Weapon, Cartridge for, Blank – See *Blank Cartridge for Weapon.*

Weapon, Explosive – Explosive Weapon: *Any explosive or incendiary bomb, grenade, rocket, or mine that is designed, made, or adapted for the purpose of inflicting serious bodily injury, death, or substantial property damage, or for the principal purpose of causing such a loud report as to cause undue public alarm or terror, and includes a device designed, made, or adapted for delivery or shooting an explosive weapon. [*Quoted from the Texas State (USA) Penal Code 46.01(2)]

Weapon, Incendiary – Incendiary Weapon: A weapon that is designed, made, or adapted to use a combustible material and an ignition mechanism for causing bodily injury, death, or property damage by fire.

Weapon, Kiloton – Kiloton Weapon: A nuclear weapon with an explosive power equivalent to one thousand tons TNT.

Weapon, Megaton – Megaton Weapon: A nuclear weapon with an explosive power equivalent to one million tons TNT.

Compare: *Kiloton Weapon.*

Weapon, Nuclear – Nuclear weapon: A weapon that uses radioactive material and in which the explosion is caused by nuclear fission, nuclear fusion, or a combination of both.

The fission bomb (atom[ic] bomb or A-bomb) consists essentially of two or more masses of a suitable fissile material (such as uranium-235 or plutonium-239), each of which is less than the critical mass. When the bomb is detonated the sub-critical masses are brought rapidly together to form a supercritical assembly, so that a single fission at the instant of contact sets off an uncontrolled chain reaction.

The resulting release of nuclear energy produces a devastating explosion, the effect of which is comparable to the explosion of tens of kilotons of T.N.T. with temperatures of the order 10^8 K being reached.

The fusion bomb (thermonuclear weapon, hydrogen bomb, or H-bomb) relies on a nuclear-fusion reaction, which becomes self-sustaining at a critical temperature of about 35×10^6 K.

Hydrogen bombs consist of either two-phase fission-fusion devices in which an inner fission bomb is surrounded by a hydrogenous material, such as heavy hydrogen (deuterium) or lithium deuteride, or a three-phase fission-fusion-fission device, which is even more powerful. The megaton explosion produced by such a thermonuclear reaction has not yet been used in war. A special type of fission-fusion bomb is called a neutron bomb, in which most of the energy is released as high-energy neutrons. This neutron radiation destroys people but provides less of the shock waves and blast that destroy buildings.

Weapon of Mass Destruction – Nuclear, biological, and chemical weapons are known as weapons of mass destruction.

Web Thickness – The distance of travel of a burning surface in a propellant grain to give complete combustion.

Wedge Cut – A method of tunneling, etc., in which at the outset the blowing out of a wedge of rock produces a free face.

Weeping – The separation of oily ingredients out of explosives during prolonged storage. In nitroglycerine gelatin explosives, weeping means liberation of nitroglycerine following breakdown of the gelatinous base.

Also see: *Exudation.*

Weight Strength – The energy of an explosive material per unit of weight expressed as a percentage of the energy per unit of weight of *straight* *nitroglycerin dynamite*, *blasting gelatin* or other explosive standard. ANFO is typically the standard these days (relative). Absolute strength is given in energy/mass units.

Weight Strength, Absolute – Absolute Weight Strength (AWS): The heat of reaction available in each gram of explosive.

Also see: *Explosive Energy.*

Weight Strength, Relative – Relative Weight Strength (RWS): The heat of reaction per unit weight of an explosive compared to the energy of an equal weight of standard ANFO.

Also see: *Explosive Energy.*

Welding – Joining of two metal surfaces by raising their temperature sufficiently to melt and fuse them together.

Welding, Explosion – Explosion Welding: The metalworking technique that uses controlled detonations to force dissimilar metal plates into a high-quality, metallurgically bonded joint. It is a solid state welding process. Such welding is also considered a cold-welding process that allows metals to be joined without losing their pre-bonded properties.

The transition joint can withstand drastic thermal excursions, is ultra-high vacuum tight, and has high mechanical strength. It can introduce thin, diffusion inhibiting inter layers such as tantalum and titanium, which allow conventional weld-up installation.

The discovery of explosion bonding process was interesting. It was discovered through the use of artillery. Varieties of driving bands and seals have been used over the years to improve the fit of artillery shells in the bore of the gun. Often this was in the form of a plate. Due to its mildness copper plates were used to preserve the bore of the gun. However, it was discovered that the plates became permanently fused to the projectile after firing. The process is capable of bonding metals together that are metallurgically incompatible. Metal blanks for sandwich coins, such as in the USA, are made this way.

Explosion bonded joints are applied anywhere a designer needs to make a high-quality transition between metals. One especially valuable feature of explosion bonding is that it can frequently be applied to metallurgically incompatible metals such as aluminum and steel or titanium and steel. Typical uses include ultra-high vacuum joints between aluminum, copper and stainless steel, corrosion resistant claddings on mild steel substrates, and alloy aluminum joined to low-expansion rate metals for electronic packages. Difficult metals such as beryllium, Al-Be alloys and rhenium can be joined with explosion welding. Powder metal products such as Al-SiC can be joined to wrought metal without thermal excursions. Stainless steel is often joined to ordinary steel in this manner.

Synonyms: Explosion Bonding, and Explosion Cladding.

Welding, Explosive – Explosive Welding: A welding done by the force of a controlled explosion.

WhC – White Compound: 1, 9-dicarboxy 2, 4, 6, 8-tetranitrophenazin-N-oxide.

Wheel, Catherine – Catherine Wheel: A circular-shaped firework.

Synonym: Pinwheel.

Wheel, Pin – Pinwheel: Synonym of *Catherine Wheel.*

White Phosphorus – The element phosphorus in its allotropic form. White phosphorus is a yellow, soft, waxy, poisonous solid, which ignites spontaneously when exposed to air. Its melting point is 44.5°C. It is soluble in carbon disulfide, insoluble in water and alcohol.

Synonym: Yellow Phosphorus.

Whizbang – 1. A small high-velocity shell which makes a whizzing sound followed by a bang when it hits.

2. A firecracker that like the whizbang shell makes a whizzing sound followed by a loud explosion.

Synonym: Whizzbang.

Wind, Fire – See *Fire Wind.*

Windshield – Ballistic Cap.

Wire, Blasting, Permanent – See *Permanent Blasting Wire.*

Wire, Bridge – See *Bridge Wire.*

Wire, Bridge, Exploding – Exploding Bridge Wire: A bridge wire designed and used to be exploded by a high-energy discharge rather than being heated by applied power. A fine metallic wire (typically gold, 1mm in length, 0.038 mm in diameter) that explodes, rather than merely heats, and also produces a shockwave, upon application of a high current through the device bridge wire. The explosion cannot be initiated by any normal shock or electrical energy. It is initiated by capacitor discharge. The wire takes the place of the primary explosive in an electric detonator.

Wire, Bus – Bus Wire: Copper wire, which is expendable, heavy-gauge and bare, used to connect detonators or series of detonators in parallel.

Wire, Lead – Lead Wire: The line (wire) connecting the electrical power source with the electric detonator circuit.

Synonym: Firing line, Leading Line, Lead Line, and Leading Wire.

Wire, Leading – Leading Wire: See *Wire, Lead***.**

Wire, Leg – Leg Wire: Leg wires mean the two single wires or one duplex wire extending out from an electric detonator, which are permanently attached to the electric detonator.

Wire, Priming – Priming Wire: A pointed wire used to penetrate the vent of a piece, for piercing the cartridge before priming.

Wire, Safety – Safety Wire: Wire set into the body of a fuse to lock all movable parts into safe position so that the fuse is not set off accidentally. Safety wire is pulled out just before loading.

Wooden Kitchen Match – Synonym of *Strike Anywhere Match.*

Wool, collodion – Collodion Wool: Synonym of *Cellulose Nitrate.*

Work, Hot – Hot Work: Any operation that involves cutting or welding by gas or electric arc; or the use of a mechanical tool that can produce sparks; or any non-welding work of equivalent risk, such as grinding, drilling, or the use of percussion tools.

Wulff Process – A chemical process to make acetylene and ethylene by cracking a hydrocarbon gas, such as butane, with high-temperature steam in a regenerative furnace.

X-310 A – An *igniter* for mild *detonating fuse.*

X-Stoff – German version of Tetranitromethane, TNM. The oxygen-rich derivative of methane is not truly an explosive by itself, but forms highly brisant mixtures with hydrocarbon, such as toluene.

Xilit – "Trinitroxylene" in Russian.

XTX – *RDX* coated with a silicon resin, namely Sylgard.

Xyloidine – Xyloidine is a synonym of *Nitrostarch,* $[C_6H_7O_2(ONO_2)_3]n$. It is not a definite compound, but a mixture of various nitric acid esters of starch of different degrees of nitration. Nitrostarch, with nitrogen contents varying from 12 to 13.3%, is prepared by nitration of starch. The empirical formula of the structural unit of nitro starch is $C_6H_7N_3O_9$. It is a pale yellow powder. It is insoluble in water and unlike starch it does not gelatinize when heated with water. It dissolves in acetone and alcohol-ether mixtures.

Synonyms: *Starch Nitrate.*

Xylene – Xylene, $C_6H_4\ (CH_3)_2$ is any one of the family of isomeric, colorless aromatic hydrocarbon liquids, produced by the destructive distillation of coal or by the catalytic reforming of petroleum naphthenic fractions. It is used for high-octane and aviation gasoline, solvents, chemical intermediates, and the manufacture of polyester resins.

Synonyms: Dimethylbenzene, and Xylol

Xylol – Synonym of *Dimethylbenzene,* and *Xylene.*

Yellow Phosphorus – Synonym of white phosphorous. It is the element phosphorus in its allotropic form. It is a yellow, soft, waxy, poisonous solid, which ignites spontaneously when exposed to air. Its melting point is at 44.5°C; is soluble in carbon disulfide, insoluble in water and alcohol.

It is used as a filling for various projectiles as a smoke-producing agent, and has an incendiary effect. It may be mixed with a xylene solution of synthetic rubber to form plasticized white phosphorous.

Synonym: *White Phosphorus.*

Yellow Salt – The chemical name of yellow salt is Uranium Nitrate, $UO_2(NO_3)_2.6H_2O$. It is an unstable yellow crystalline salt, which melts at 60°C and boils at 118°C. It has toxic and explosive characteristics. It is soluble in water, alcohol, and ether. Yellow salt is used in photography, in medicine, and for uranium extraction and uranium glaze.

Synonyms: Uranium Nitrate, and Uranyl Nitrate.

Yonkite – "Industrial explosive" in Belgian.

Yperite – French name for Dichlorodiethyl Sulphide, $(CH_2CH_2Cl)_2S$. It is an oily liquid, which has been used as a war gas.

Yuenyaku – Japanese name for *Black Powder.*

Zapon – A solution of *nitrocellulose* used in the manufacture of fuse head.

Zazhigateinaya – Russian synonym for *Molotov cocktail. Zazhigateinaya* means an explosive liquid.

Zero Delay Detonator – It is a *detonator* that has a firing time of essentially zero seconds in comparison to delay detonators with firing times of several milliseconds to several seconds. In other words, it is a detonator that contains no delay element. A zero delay detonator may be either electric or non-electric.

However, the nomenclature 'zero delay' (instantaneous) is somewhat misleading as the function times can be as long as 5 milliseconds after the application of the initiating energy. The function time of zero delay non-electric detonators must include the travel time of the initiating signal through the cord or shock tubing.

Synonym: *Instantaneous Detonator.*

Zinc Peroxide – Light yellow amorphous powder. It is used in pyrotechnic and primer compositions.

Zirconium – An element (Zr) having atomic number 40, specific gravity 6.5 and melting point 1800°C. The element has hard, lustrous, grayish, crystalline scales. A finely divided form of Zirconium is used as oxygen-acceptor in priming mixtures. It is used in fuse heads in combination with lead styphnate.

Zirconium Picramate – Zirconium picramate when dry or containing, by weight, less than 20% water is an explosive having a fire hazard with minor or no explosion effects. That is to say, it is an explosive belonging to division 1.3.

Zirconium picramate, when wetted with not less than 20% water, falls under the class 5.1 dangerous substance. This is an oxidizing substance which while in itself is not necessarily combustible, may, generally yielding oxygen, cause or contribute to the combustion of other material.

Zirconium picramate in wet condition is a slurry or sludge of yellow crystals. It may take some effort to ignite but once ignited the material will burn with great rapidity. It will accelerate the burning of combustible materials. If large quantities are involved in fire or if containers of the material are exposed to fire or heat for prolonged periods, an explosion may result. Toxic oxides of nitrogen are produced during combustion.

Zone, Cold – Cold Zone: The area where equipment and personnel directly support the incident.

Synonyms: Safe Zone, and Support Zone.

Zone, Danger – Danger Zone: The area within which persons would be in physical jeopardy due to overpressure, fragments or firebrands released during shot or round of shots.

Zone, Fizz – Fizz Zone: In the burning of a propellant, the zone in which the solid propellant is converted to gaseous intermediates.

Zone, Detonation – Detonation Zone: In the *detonation* of an explosive, that portion of the *detonating explosive* in which the chemical reaction is taking place comprises a chemical reaction zone preceded by a *shock wave.*

Zone, Flame – Flame Zone: The final stage, in the burning of a propellant, in which gaseous intermediates react with the production of a flame.

Zone, Foam – Foam Zone: The initial stage of partial gasification in the burning of a propellant.

Zone, Hot – Hot Zone: The zone immediately surrounding the chemical or biological release. This zone extends far enough to prevent adverse effects to personnel.

Zone, Safe – Safe Zone: Synonym of *Cold Zone.*

See *Zone, Cold.*

Zone, Safety – Safety Zone: The minimum linear distance around an explosive magazine, a pressure vessel or a storage shed/tank of inflammable liquid which is maintained unoccupied so that in case of an explosion or fire the least possible damage will be done to the protected works, personnel or environment in the surrounding areas. Safety distance is set by hazard calculation followed by any practicing convention.

Synonym: Safety Distance.

Zone, Support – Support Zone: Synonym of *Cold Zone*.

Zyklon B – A chemical warfare agent, which is a variation of hydrogen cyanide. It is a powder, extremely volatile and toxic, and can be dispersed over a large area by placing it into fume.

Zylonite – Synonym of *Celluloid.*

Bibliography

ISEE Blasters Handbook, International Society of Explosives Engineers, USA.

Blasting Explosives: A Guide to Safety, Explosives Regulatory Division, Natural Resources Canada.

Blasting Explosives and Detonators: Storage, possession, Transportation, Distribution and Sale, Explosives Regulatory Division, Natural Resources Canada.

Blasting Practice, Nobel Division, Imperial Chemical Industries (ICI), UK.

Blasting Primer: A Study Guide for Students of Explosives Engineering, James T. Ludwiczak (Blasting and Mining Consultants, Inc. KY).

Blasting Technology for Mining and Civil Engineers, Gour C. Sen (University of New South Wales Press Ltd, Australia).

The Chemistry and Characteristics of Explosive Material, James R. Cook (Vantage Press New York)

The Chemistry of Explosives, J Akhavan (The Royal Society of Chemistry, UK).

The Chemistry of Powder & Explosives, Tenney L Davis (Angriff Press, Nevada, USA).

Dangerous Properties of Industrial Materials, N. Irving Sax (Reinold Publishing House, New York).

Detonation Theory and Experiment, Wildon Fickett and William C. Davis (Dover Publications, Inc., New York).

Dictionary of Chemistry, Penguin Books, UK.

Dictionary of Physics, Oxford University Press, UK.

Dictionary of Science, E.B. Uvarov and Alan Isaacs (Penguin Books Ltd, UK).

Explosion Prevention Systems, NFPA 69 (National Fire Protection Association, USA).

Explosives, Rudolf Meyer, Josef Kohler, Axel Homburg (Wiley-Vch Verlag GmbH, Weinheim, Germany).

Explosives, A. Marshal (J. & A. Churchill, London).

Explosives: An Engineering Tool, Giorgio Berta (Italesplosivi-Milano).

Explosives Engineering, Paul W. cooper, (Wiley-Vch, New York).

Explosives Science, Raymond N. Rogers and Joan L. Rogers (Los Alamos, NM, USA).

General Fire Hazards and Fire Prevention, J.J. Williamson (Sir Isaac Pitman & Sons Ltd., UK).

High Explosives and Propellants, S. Fordham (Pergamon Press, UK).

The Journal of **Explosives Engineering,** International Society of Explosives Engineers, USA.

Major Hazard Control: A Practical Manual, ILO, Geneva.

Manual of Explosives Military Pyrotechnics and Chemical Warfare Agents, Jules Bebie (The Macmillan Company, New York).

Manufacture, Transportation, Storage and Use of Explosives Materials, NFPA 495 (National Fire Protection Association, USA).

Preparing for Terrorism: An Emergency Services Guide, George Buck (Delmar Publishers, USA).

Protection Against Bombs and Incendiaries, Earl A. Pike (Charles C Thomas Publisher, USA).

IME Safety Library Publications Nos. 1-22, Institute of Makers of Explosives, USA.

Storage Standards for Industrial Explosives, Explosives Regulatory Division, Natural Resources Canada.

Terrorism Handbook for Operational Responders, A Bevelacqua, R. Stilp (Delmar Publishers, USA).

Test Methods for Explosives, Muhamed Suceska (Springer-Verlag, New York, Inc. NY).

Technical Bulletin of the Blasting Products Division, The Ensign-Bickford Company, USA.

Transport of Dangerous Goods (Orange Book): Recommendations of the United Nations Committee of Experts (UN Code).